T0094103

Springer Optimization and Its Applications 63

Series Editor:
Panos M. Pardalos

Subseries:
Nonconvex Optimization and Its Applications

For further volumes:
http://www.springer.com/series/7393

Daniel Scholz

Deterministic Global Optimization

Geometric Branch-and-bound Methods and Their Applications

Daniel Scholz
Institute for Numerical and Applied Mathematics
Georg-August-University Göttingen
Lotzestraße 16-18
37083 Göttingen
Germany
dscholz@math.uni-goettingen.de

ISSN 1931-6828
ISBN 978-1-4614-1950-1 e-ISBN 978-1-4614-1951-8
DOI 10.1007/978-1-4614-1951-8
Springer New York Dordrecht Heidelberg London

Library of Congress Control Number: 2011941012

Mathematics Subject Classification (2010): 90C26, 90C30, 90C29

Printed on acid-free paper

Springer is part of Springer Science+Business Media (www.springer.com)

Preface

Almost all areas of science, economics, and engineering rely on optimization problems where global optimal solutions have to be found; that is one wants to find the global minima of some real-valued functions. But because in general several local optima exist, global optimal solutions cannot be found by classical nonlinear programming techniques such as convex optimization. Hence, ***deterministic global optimization*** comes into play. Applications of deterministic global optimization problems can be found, for example, in computational biology, computer science, operations research, and engineering design among many other areas.

Many new theoretical and computational contributions to deterministic global optimization have been developed in the last decades and ***geometric branch-and-bound methods*** arose to a commonly used solution technique. The main task throughout these algorithms is to calculate lower bounds on the objective function and several methods to do so can be found in the literature. All these techniques were developed in parallel, therefore the main contribution of the present book is a general theory for the evaluation of bounding operations, namely the ***rate of convergence***. Furthermore, several extensions of the basic prototype algorithm as well as some applications of geometric branch-and-bound methods can be found in the following chapters. We remark that our results are restricted to unconstrained global optimization problems although constrained problems can also be solved by geometric branch-and-bound methods using the same techniques for the calculation of lower bounds.

All theoretical findings in this book are evaluated numerically. To this end, we mainly make use of continuous ***location theory***, an area of operations research where geometric branch-and-bound methods are suitable solution techniques. Our computer programs were coded in Java using double precision arithmetic and all tests were run on a standard personal computer with 2.4 GHz and 4 GB of memory. Note that all programs were not optimized in their runtimes, i.e., some more efficient implementations might be possible.

The chapters in the present book are divided into three parts. The prototype algorithm and its bounding operations can be found in the first part in Chapters 1 to 3. Some problem extensions are discussed in the second part; see Chapters 4 to 6. Finally, the third part deals with applications given in Chapters 7 to 9. A suggested order of reading can be found in Figure 0.1.

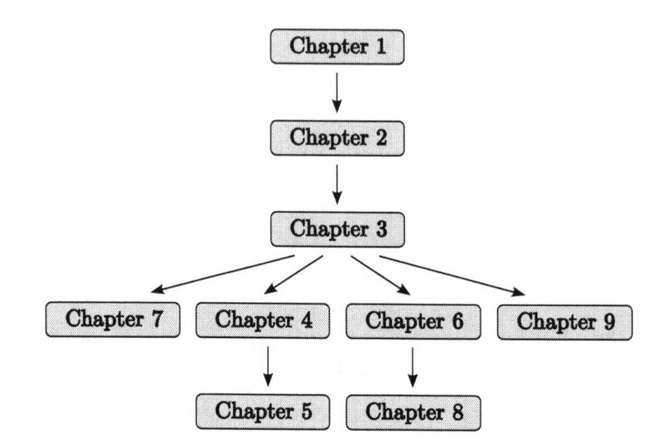

Fig. 0.1 Suggested order of reading.

In detail, the remainder is structured as follows.

In ***Chapter 1***, we present some preliminaries that are important for the understanding of the following chapters. We recall the definition and basic results of convex functions and generalizations of convexity are discussed. Next, we give a brief introduction to location theory before the class of d.c. functions and its algebra are presented. The chapter ends with an introduction to interval analysis.

The geometric branch-and-bound prototype algorithm is introduced in ***Chapter 2***. Here, we start with a literature review before we introduce the definition of bounding operations. Finally, we suggest a definition for the rate of convergence which leads to the most important definition in the following chapter and to a general convergence theory.

The main contribution of the present work can be found in ***Chapter 3***. Therein, we make use of the suggested rate of convergence which is discussed for nine bounding operations, among them some known ones from the literature as well as some new bounding procedures. In all cases, we prove the theoretical rate of convergence. Furthermore, some numerical results justify our theoretical findings and the empirical rate of convergence is computed.

In ***Chapter 4*** we introduce the first extension of the branch-and-bound algorithm, namely an extension to multicriteria problems. To this end, we briefly summarize the basic ideas of multicriteria optimization problems before the algorithm is suggested. Moreover, we present a general convergence theory as well as some numerical examples on two bicriteria facility location problems.

The multicriteria branch-and-bound method is further extended in ***Chapter 5*** where some general discarding tests are introduced. To be more precise, we make use of necessary conditions for Pareto optimality such that the algorithm results in a very sharp outer approximation of the set of all Pareto optimal solutions. The theoretical findings are again evaluated on some facility location problems.

A third extension can be found in ***Chapter 6***. Therein, we assume that the objective function does not only depend on continuous variables but also on some combinatorial ones. Here, we generalize the concept of the rate of convergence and some bounding operations are suggested. Furthermore, under certain conditions this extension leads to exact optimal solutions as also shown by some location problems on the plane.

In the following ***Chapter 7*** we present a first application of the geometric branch-and-bound method, namely the circle detection problem. We show how global optimization techniques can be used to detect shapes such as lines, circles, and ellipses in images. To this end, we discuss the general problem formulation before lower bounds for the circle detection problem are given. Some numerical results show that the method is highly accurate.

A second application can be found in ***Chapter 8*** where integrated scheduling and location problems are discussed. After an introduction to the planar ScheLoc makespan problem we mainly make use of our results from Chapter 6. We derive lower bounds and show how to compute an exact optimal solution. The numerical results show that the proposed method is much faster than other techniques reported in the literature.

Another interesting location problem is presented in ***Chapter 9***, namely the median line location problem in three-dimensional Euclidean space. Some theoretical results as well as a specific four-dimensional problem parameterization are discussed before we suggest some lower bounds using the techniques given in Chapter 3. Moreover, we show how to find an initial box that contains at least one optimal solution.

Finally, we conclude our work with a summary and a discussion in ***Chapter 10***. In addition, some extensions and ideas for further research are given.

Acknowledgments

The present book is a joint work with several coworkers who shared some research experience with me during the last several years. First of all I would like to thank Anita Schöbel for fruitful discussions, enduring support, and her pleasant and constructive co-operation. Several chapters of this book were written in close collaboration with her. Moreover, I thank Emilio Carrizosa and Rafael Blanquero for our joint work in Chapter 9, Marcel Kalsch for the collaboration in ScheLoc problems, and Hauke Strasdat for several discussions on Chapter 7.

Next, I would like to thank everybody else who supported me during my research activities in recent years. In particular, my thanks go to Michael Weyrauch for our joint work in mathematical physics and to my colleagues in Göttingen, Annika Eickhoff-Schachtebeck, Marc Goerigk, Mark-Christoph Körner, Thorsten Krempasky, Michael Schachtebeck, and Marie Schmidt. I very much enjoyed the time with you in and outside the office. Special thanks go to Michael Schachtebeck for proof reading the manuscript and for his helpful comments.

Finally, I thank all people the who supported me in any academic or nonacademic matters. In particular, I want to thank my family for their enduring support and patience.

Göttingen, February 2011 *Daniel Scholz*

Contents

Symbols and notations

Important variables

n	dimension of the domain of the objective function $f : \mathbb{R}^n \to \mathbb{R}$
m	number of demand points
p	number of objective functions in multicriteria optimization problems
s	number of subboxes generated in each step: Y is split into Y_1 to Y_s

Superscripts

L	left endpoint of an interval
R	right endpoint of an interval
T	transposed vector

Multicriteria optimization

$x \leqq y$	if $x = (x_1, \ldots, x_n), y = (y_1, \ldots, y_n) \in \mathbb{R}^n$, and $x_k \leq y_k$ for $k = 1, \ldots, n$
$x \leq y$	if $x = (x_1, \ldots, x_n), y = (y_1, \ldots, y_n) \in \mathbb{R}^n$, and $x \leqq y$ with $x \neq y$
$x < y$	if $x = (x_1, \ldots, x_n), y = (y_1, \ldots, y_n) \in \mathbb{R}^n$, and $x_k < y_k$ for $k = 1, \ldots, n$
$\mathbb{R}^p_{\geqq}$	symbol for the set $\{x \in \mathbb{R}^p : x \geqq 0\}$
$\mathbb{R}^p_{\geq}$	symbol for the set $\{x \in \mathbb{R}^p : x \geq 0\}$
$\mathbb{R}^p_{>}$	symbol for the set $\{x \in \mathbb{R}^p : x > 0\}$
X_E	set of all Pareto optimal solutions: $X_E \subset \mathbb{R}^n$
Y_N	set of all nondominated points: $Y_n \subset \mathbb{R}^p$
X_{wE}	set of all weakly Pareto optimal solutions: $X_{wE} \subset \mathbb{R}^n$
Y_{wN}	set of all weakly nondominated points: $Y_{wN} \subset \mathbb{R}^p$
X_E^ε	set of all ε-Pareto optimal solutions: $X_E^\varepsilon \subset \mathbb{R}^n$
Y_N^ε	set of all ε-nondominated points: $Y_N^\varepsilon \subset \mathbb{R}^p$
X_A^ε	output set of the multicriteria branch-and-bound method: $X_A^\varepsilon \subset \mathbb{R}^n$

Bounding operations

$c(Y)$	center of a box $Y \subset \mathbb{R}^n$
$\delta(Y)$	Euclidean diameter of a box $Y \subset \mathbb{R}^n$
$V(Y)$	set of the 2^n vertices of a box $Y \subset \mathbb{R}^n$
$\Omega(Y)$	dominating set for mixed combinatorial problems for a box $Y \subset \mathbb{R}^n$
M	threshold for the cardinality of $\Omega(Y)$

Miscellaneous

(a,b)	open interval
$[a,b]$	closed interval
$A \subset B$	a set A is included in or equal to a set B
$\lceil \alpha \rceil$	smallest integer greater than or equal to $\alpha \in \mathbb{R}$
$\lfloor \alpha \rfloor$	greatest integer less than or equal to $\alpha \in \mathbb{R}$
$\nabla f(c)$	gradient of $f : \mathbb{R}^n \to \mathbb{R}$ at $c \in \mathbb{R}^n$
$Df(c)$	Jacobian matrix of $f : \mathbb{R}^n \to \mathbb{R}^p$ at $c \in \mathbb{R}^n$
$D^2 f(c)$	Hessian matrix of $f : \mathbb{R}^n \to \mathbb{R}$ at $c \in \mathbb{R}^n$
$\partial f(b)$	subdifferential of a convex function $f : \mathbb{R}^n \to \mathbb{R}$ at $b \in \mathbb{R}^n$
$\lvert X \rvert$	cardinality of a (finite) set X
$\lVert x \rVert_1$	rectilinear norm of $x \in \mathbb{R}^n$
$\lVert x \rVert_2$	Euclidean norm of $x \in \mathbb{R}^n$
$f \circ g$	composition of two functions: $(f \circ g)(x) = f(g(x))$
$\text{conv}(A)$	convex hull of $A \subset \mathbb{R}^n$
Π_m	set of all permutations of length m

Chapter 1
Principles and basic concepts

Abstract In this chapter, our main goal is to summarize principles and basic concepts that are of fundamental importance in the remainder of this text, especially in Chapter 3 where bounding operations are presented. We begin with the definition of convex functions and some generalizations of convexity in Section 1.1. Some fundamental but important results are given before we discuss subgradients. Next, in Section 1.2 we briefly introduce distance measures given by norms. Distance measures are quite important in the subsequent Section 1.3 where we give a very brief introduction to location theory. Furthermore, we show how to solve the Weber problem for the rectilinear and Euclidean norms. Moreover, d.c. functions are introduced in Section 1.4 and basic properties are collected. Finally, we give an introduction to interval analysis in Section 1.5 which leads to several bounding operations later on.

1.1 Convex functions and subgradients

One of the most important and fundamental concepts in this work is convex and concave functions.

Definition 1.1. A set $X \subset \mathbb{R}^n$ is called ***convex*** if

$$\lambda \cdot x + (1-\lambda) \cdot y \in X$$

for all $x, y \in X$ and all $\lambda \in [0,1]$.

Definition 1.2. Let $X \subset \mathbb{R}^n$ be a convex set. A function $f : X \to \mathbb{R}$ is called ***convex*** if

$$f(\lambda \cdot x + (1-\lambda) \cdot y) \le \lambda \cdot f(x) + (1-\lambda) \cdot f(y)$$

for all $x, y \in X$ and all $\lambda \in [0,1]$. A function f is called ***concave*** if $-f$ is convex; that is if

$$f(\lambda \cdot x + (1-\lambda) \cdot y) \ge \lambda \cdot f(x) + (1-\lambda) \cdot f(y)$$

for all $x, y \in X$ and all $\lambda \in [0,1]$.

Convex functions have the following property.

Theorem 1.1. *Let $X \subset \mathbb{R}^n$ be a convex set. Then a twice differentiable function $f : X \to \mathbb{R}$ is convex if and only if the Hessian matrix $D^2 f(x)$ is positive semidefinite for all x in the interior of X.*

Proof. See, for instance, Rockafellar (1970). □

Several generalizations of convex and concave functions can be found, for example, in Avriel et al. (1987), among them quasiconcave functions. This property is important for the geometric branch-and-bound algorithm.

Definition 1.3. Let $X \subset \mathbb{R}^n$ be a convex set. A function $f : X \to \mathbb{R}$ is called ***quasiconvex*** if

$$f(\lambda \cdot x + (1-\lambda) \cdot y) \leq \max\{f(x), f(y)\}$$

for all $x, y \in X$ and all $\lambda \in [0,1]$. A function f is called ***quasiconcave*** if $-f$ is quasiconvex; that is if

$$f(\lambda \cdot x + (1-\lambda) \cdot y) \geq \min\{f(x), f(y)\}$$

for all $x, y \in X$ and all $\lambda \in [0,1]$.

Lemma 1.1. *Every convex function f is quasiconvex and every concave function f is quasiconcave.*

Proof. Let f be convex. Then we obtain

$$\begin{aligned} f(\lambda \cdot x + (1-\lambda) \cdot y) &\leq \lambda \cdot f(x) + (1-\lambda) \cdot f(y) \\ &\leq \lambda \cdot \max\{f(x), f(y)\} + (1-\lambda) \cdot \max\{f(x), f(y)\} \\ &= \max\{f(x), f(y)\}. \end{aligned}$$

Hence, f is quasiconvex. In the same way it can be shown that every concave function f is quasiconcave. □

Note that the reverse of Lemma 1.1 does not hold.

Lemma 1.2. *Let $X \subset \mathbb{R}^n$ be a convex set and consider a concave function $g : X \to \mathbb{R}$ with $g(x) \geq 0$ and a convex function $h : X \to \mathbb{R}$ with $h(x) > 0$ for all $x \in X$. Then $f : X \to \mathbb{R}$ defined for all $x \in X$ by*

$$f(x) := \frac{g(x)}{h(x)}$$

is quasiconcave.

Proof. See, for instance, Avriel et al. (1987). □

Moreover, from the definition of convexity we directly obtain the following result.

Lemma 1.3. *Let $X \subset \mathbb{R}^n$ be a convex set, let $\lambda, \mu \geq 0$, and consider two convex (concave) functions $g, h : X \to \mathbb{R}$. Then $\lambda g + \mu h$ is also a convex (concave) function.*

Note that this result does not hold for quasiconvex (quasiconcave) functions; that is the sum of quasiconvex (quasiconcave) might not be quasiconvex (quasiconcave) any more.

Definition 1.4. Let $a_1, \ldots, a_m \in \mathbb{R}^n$ be a finite set of points. Then the ***convex hull*** of these points is defined as

$$\mathrm{conv}(a_1, \ldots, a_m) := \left\{ \sum_{k=1}^{m} \lambda_k a_k \ : \ \lambda_1, \ldots, \lambda_m \geq 0, \ \lambda_1 + \cdots + \lambda_m = 1 \right\}.$$

The convex hull $\mathrm{conv}(a_1, \ldots, a_m)$ is a convex set.

Example 1.1. Consider the set

$$X = [x_1^L, x_1^R] \times \cdots \times [x_n^L, x_n^R] \subset \mathbb{R}^n$$

and let $\{a_1, \ldots, a_{2^n}\}$ be the 2^n vertices of X. Then we obtain

$$\mathrm{conv}(a_1, \ldots, a_{2^n}) = X.$$

The following important result says that a minimum of a concave or quasiconcave function over a convex hull of a finite set of points can be computed easily.

Lemma 1.4. *Let $X = \mathrm{conv}(a_1, \ldots, a_m)$ be the convex hull of $a_1, \ldots, a_m \in \mathbb{R}^n$ and consider a concave or quasiconcave function $f : \mathbb{R}^n \to \mathbb{R}$. Then*

$$\min_{x \in X} f(x) = \min\{f(a_k) \ : \ k = 1, \ldots, m\}.$$

Proof. See, for instance, Horst and Tuy (1996). □

Finally, we need the concept of subgradients for convex functions; see, for example, Rockafellar (1970) or Hiriart-Urruty and Lemaréchal (2004).

Definition 1.5. Let $X \subset \mathbb{R}^n$ be a convex set and let $f : X \to \mathbb{R}$ be a convex function. A vector $\xi \in \mathbb{R}^n$ is called a ***subgradient*** of f at $b \in X$ if

$$f(x) \geq f(b) + \xi^T (x - b) \ \text{ for all } x \in X.$$

The set of all subgradients of f at b is called the ***subdifferential*** of f at b and is denoted by $\partial f(b)$.

Note that if ξ is a subgradient of f at b, then the affine linear function

$$h(x) := f(b) + \xi^T (x-b)$$

is a supporting hyperplane of f at b; that is one has $h(x) \leq f(x)$ for all $x \in X$. Furthermore, the following three results can be found, for instance, in Rockafellar (1970) and Hiriart-Urruty and Lemaréchal (2004).

Lemma 1.5. *Let $X \subset \mathbb{R}^n$ be a convex set and let $f : X \to \mathbb{R}$ be a convex function. Then there exists a subgradient of f at b for any b in the interior of X.*

Lemma 1.6. *Let $X \subset \mathbb{R}^n$ be a convex set and let $f : X \to \mathbb{R}$ be a convex function. Then the subdifferential of f at b is a convex set for any $b \in X$.*

Lemma 1.7. *Let $X \subset \mathbb{R}^n$ be a convex set and let $f : X \to \mathbb{R}$ be a convex function. If f is differentiable at b in the interior of X then we find*

$$\partial f(b) = \{\nabla f(b)\};$$

that is the gradient of f at b is the unique subgradient of f at b.

1.2 Distance measures

In facility location problems one wants to find a new location that, for instance, minimizes the sum of some distances to existing demand points. In all our examples, we consider norms as distance functions.

Definition 1.6. A ***norm*** is a function $\|\cdot\| : \mathbb{R}^n \to \mathbb{R}$ with the following properties.

(1) $\|x\| = 0$ if and only if $x = 0$.
(2) $\|\lambda \cdot x\| = |\lambda| \cdot \|x\|$ for all $x \in \mathbb{R}^n$ and all $\lambda \in \mathbb{R}$.
(3) $\|x+y\| \leq \|x\| + \|y\|$ for all $x, y \in \mathbb{R}^n$.

For any norm $\|\cdot\|$, the distance between two points $x, y \in \mathbb{R}^n$ is given by

$$\|x-y\| \in \mathbb{R}.$$

Example 1.2. Let $x = (x_1, \ldots, x_n) \in \mathbb{R}^n$ and $1 < p < \infty$. Some of the most important norms are the following ones.

$$\|x\|_1 := |x_1| + \cdots + |x_n|,$$

$$\|x\|_2 := \sqrt{x_1^2 + \cdots + x_n^2},$$

$$\|x\|_\infty := \max\{|x_k| \ : \ k = 1, \ldots, n\},$$

$$\|x\|_p := \left(|x_1|^p + \cdots + |x_n|^p\right)^{1/p}.$$

We call $\|\cdot\|_1$ the ***rectilinear norm***, $\|\cdot\|_2$ the ***Euclidean norm***, $\|\cdot\|_\infty$ the ***maximum norm***, and $\|\cdot\|_p$ the ***ℓ_p-norm***.

Lemma 1.8. *Let $\|\cdot\| : \mathbb{R}^n \to \mathbb{R}$ be a norm. Then $\|\cdot\|$ is a convex function.*

Proof. For all $\lambda \in [0,1]$ and $x,y \in \mathbb{R}^n$ we directly obtain

$$\|\lambda \cdot x + (1-\lambda)\cdot y\| \;\leq\; \|\lambda \cdot x\| + \|(1-\lambda)\cdot y\| \;=\; \lambda \cdot \|x\| + (1-\lambda)\cdot \|y\|$$

only using the properties (2) and (3). □

Finally, we define Lipschitzian functions using the Euclidean norm.

Definition 1.7. Let $X \subset \mathbb{R}^n$. A function $f : X \to \mathbb{R}$ is called a ***Lipschitzian function*** on X with ***Lipschitzian constant*** $L > 0$ if

$$|f(x) - f(y)| \;\leq\; L \cdot \|x-y\|_2 \;\text{ for all } x,y \in X.$$

1.3 Location theory

In classic location theory we have as given a finite set of existing demand points on the plane with weights representing the advantage or disadvantage of each demand point. The problem is to find a new facility location with respect to the given demand points, for example, a new location that minimizes the weighted sum or the weighted maximum of distances between the demand points and the new facility location. An overview of facility location problems can be found in Love et al., Drezner (1995), or Drezner and Hamacher (2001).

Although a wide range of facility location problems can be formulated as global optimization problems in small dimension, they are often hard to solve. However, geometric branch-and-bound methods are convenient and commonly used solution algorithms for these problems; see Chapter 2. Therefore, all of our algorithms in the following chapters are demonstrated on some facility location problems.

In this section, we want to present some well-known solution algorithms for one of the first facility location problems, namely the ***Weber problem***. To this end, assume m given demand points $a_k = (a_{k,1}, a_{k,2}) \in \mathbb{R}^2$ with weights $w_k \geq 0$ for $k = 1,\ldots,m$. The goal is to minimize the objective function $f : \mathbb{R}^2 \to \mathbb{R}$ defined by

$$f(x) \;=\; \sum_{k=1}^{m} w_k \cdot \|x - a_k\|,$$

where $\|\cdot\|$ is a given norm. Note that this objective function is convex due to Lemma 1.8. We now want to solve this problem for the rectilinear and the Euclidean norms; see Drezner et al. (2001) and references therein.

1.3.1 The Weber problem with rectilinear norm

Using the rectilinear norm, we obtain the objective function

$$\begin{aligned} f(x) = f(x_1,x_2) &= \sum_{k=1}^{m} w_k \cdot (|x_1 - a_{k,1}| + |x_2 - a_{k,2}|) \\ &= \sum_{k=1}^{m} w_k \cdot |x_1 - a_{k,1}| + \sum_{k=1}^{m} w_k \cdot |x_2 - a_{k,2}|. \end{aligned}$$

Hence, the problem can be reduced to the minimization of two objective functions with one variable each. We have to minimize two piecewise linear and convex functions $g : \mathbb{R} \to \mathbb{R}$ with

$$g(t) = \sum_{k=1}^{m} w_k \cdot |t - b_k|. \tag{1.1}$$

Moreover, denote by Π_m the set of all permutations of $\{1,\ldots,m\}$. Then we obtain the following solution; see, for instance, Drezner et al. (2001) or Hamacher (1995).

Find a $\pi = (\pi_1,\ldots,\pi_m) \in \Pi_m$ such that

$$b_{\pi_1} \leq \cdots \leq b_{\pi_m}$$

and define

$$s := \min\left\{ r \in \mathbb{N} : \sum_{k=1}^{r} w_{\pi_k} \geq \frac{1}{2} \cdot \sum_{k=1}^{m} w_k \right\}.$$

Then $t^* = b_{\pi_s}$ is a global minimum of g; see Equation (1.1).

In other words, an optimal solution for the Weber problem with rectilinear norm can be easily found by sorting the values $\{a_{1,1},\ldots,a_{m,1}\}$ and $\{a_{1,2},\ldots,a_{m,2}\}$.

1.3.2 The Weber problem with Euclidean norm

The objective function for the Euclidean norm is

$$f(x) = f(x_1,x_2) = \sum_{k=1}^{m} w_k \cdot \sqrt{(x_1 - a_{k,1})^2 + (x_2 - a_{k,2})^2}.$$

Before we discuss the general case, one finds the following necessary and sufficient optimality conditions for the given demand points; see, for example, Drezner et al. (2001).

Let $s \in \{1,\ldots,m\}$. If

$$\left(\sum_{\substack{k=1\\k\neq s}}^{m}\frac{w_k\cdot(a_{s,1}-a_{k,1})}{\|a_s-a_k\|_2}\right)^2+\left(\sum_{\substack{k=1\\k\neq s}}^{m}\frac{w_k\cdot(a_{s,2}-a_{k,2})}{\|a_s-a_k\|_2}\right)^2\leq w_s^2,$$

then $x^*=a_s$ is an optimal solution for the Weber problem with Euclidean norm.

The general solution algorithm was first suggested by Weiszfeld (1937) and is called the ***Weiszfeld algorithm***; see, for instance, Drezner et al. (2001).

Define $F:\mathbb{R}^2\to\mathbb{R}^2$ with

$$F(x)=\left(\sum_{k=1}^{m}\frac{w_k\cdot a_k}{\|x-a_k\|_2}\right)\cdot\left(\sum_{k=1}^{m}\frac{w_k}{\|x-a_k\|_2}\right)^{-1}.$$

Then, for any starting point $x_0\neq a_k$ for $k=1,\ldots,m$, the Weiszfeld algorithm is defined by

$$x_{k+1}:=F(x_k),$$

which leads to a popular solution technique for the Weber problem with Euclidean norm.

Many theoretical results and generalizations of the Weiszfeld algorithm can be found in the literature; see, for example, Drezner et al. (2001) and Plastria and Elosmani (2008).

1.4 D.c. functions

In this subsection, our aim is to sum up basic results concerning d.c. functions; see references such as Tuy (1998) or Horst and Thoai (1999).

Definition 1.8. Let $X\subset\mathbb{R}^n$ be a convex set. A function $f:X\to\mathbb{R}$ is called a ***d.c. function*** on X if there exist two convex functions $g,h:X\to\mathbb{R}$ such that

$$f(x)=g(x)-h(x)\ \text{ for all } x\in X.$$

Obviously, d.c. decompositions are not unique. For example, let $f(x)=g(x)-h(x)$ be a d.c. decomposition of f. Then

$$f(x)=(g(x)+a(x))-(h(x)+a(x))$$

is also a d.c. decomposition of f for any convex function a because the sum of convex functions is convex again. The following result shows that the algebra of d.c. functions is much more powerful than the algebra of convex or quasiconvex functions.

Lemma 1.9. *Let $X \subset \mathbb{R}^n$ be a convex set, let $f, f_1, \ldots, f_m : \mathbb{R}^n \to \mathbb{R}$ be d.c. functions on X, and let $\lambda_1, \ldots, \lambda_m \in \mathbb{R}$. Then the following functions are d.c. functions.*

$$g(x) = \sum_{k=1}^{m} \lambda_k f_k(x), \qquad g(x) = \prod_{k=1}^{m} f_k(x),$$

$$g(x) = \max_{k=1,\ldots,m} f_k(x), \quad g(x) = \min_{k=1,\ldots,m} f_k(x),$$

$$g(x) = \max\{0, f(x)\}, \quad g(x) = \min\{0, f(x)\},$$

$$g(x) = |f(x)|.$$

Proof. A constructive proof can be found in Tuy (1998). □

The following result can also be proven constructively.

Theorem 1.2. *Let $X \subset \mathbb{R}^n$ be a convex set and assume that $f : X \to \mathbb{R}$ is twice continuously differentiable on X. Then f is a d.c. function.*

Proof. See, for instance, Tuy (1998) or Horst and Thoai (1999). □

Although the previous results lead to general calculations of d.c. decomposition, there are several other methods to do so in an easier way; see, for example, Ferrer (2001) for d.c. decompositions of polynomial functions. However, in general it is far from trivial to derive a d.c. decomposition. For a further detailed discussion about d.c. decompositions we refer to Tuy (1998) and Horst and Tuy (1996).

However, the following results show how to construct suitable d.c. decompositions under certain assumptions that we use in following chapters. Although more general results can be found in Tuy (1998), we also present the proofs for our special cases.

Lemma 1.10. *Let $X \subset \mathbb{R}^n$ be a convex set and consider two convex functions*

$$r, s : X \to [0, \infty).$$

Then $r \cdot s$ is a d.c. function on X with d.c. decomposition

$$(r \cdot s)(x) = r(x) \cdot s(x) = \frac{1}{2}\Big(r(x) + s(x)\Big)^2 - \frac{1}{2}\Big(r(x)^2 + s(x)^2\Big).$$

Proof. Inasmuch as $c : [0, \infty) \to \mathbb{R}$ with $c(x) = x^2$ is increasing and convex on $[0, \infty)$, we easily find that the functions r^2, s^2, and $(r+s)^2$ are convex on X. Thus,

$$g(x) = \frac{1}{2}\Big(r(x) + s(x)\Big)^2 \text{ and } h(x) = \frac{1}{2}\Big(r(x)^2 + s(x)^2\Big)$$

is a d.c. decomposition for $r \cdot s$. □

Lemma 1.11. *Let $X \subset \mathbb{R}^n$ be a convex set and assume that the functions $s : X \to [0, \infty)$ and $r : [0, \infty) \to \mathbb{R}$ are convex and twice continuously differentiable on X*

and $[0,\infty)$*, respectively. Furthermore, assume that* r *is nonincreasing and define* $z = r'(0)$.

Then $r \circ s$ *is a d.c. function on* X *with d.c. decomposition*

$$(r \circ s)(x) \;=\; r(s(x)) \;=\; g(x) - h(x),$$

where $g(x) = r(s(x)) - z \cdot s(x)$ *and* $h(x) = -z \cdot s(x)$.

Proof. Because r is nonincreasing, we obtain that $z \le 0$. Thus, h is a convex function. Next, we calculate the Hessian matrix $D^2 g(x)$ of g for any $x \in X$:

$$\begin{aligned} D^2 g(x) &= r''(s(x)) \cdot \nabla s(x) \cdot (\nabla s(x))^T + r'(s(x)) \cdot D^2 s(x) - z \cdot D^2 s(x) \\ &= r''(s(x)) \cdot \nabla s(x) \cdot (\nabla s(x))^T + \Big(r'(s(x)) - z\Big) \cdot D^2 s(x). \end{aligned}$$

Inasmuch as r is convex, we have $r''(s(x)) \ge 0$ and

$$r'(s(x)) - z \;\ge\; r'(0) - z \;=\; 0$$

for all $x \in X$. Furthermore, $D^2 s(x)$ and $\nabla s(x) \cdot (\nabla s(x))^T$ are positive semidefinite matrices. To sum up, $D^2 g(x)$ is also positive semidefinite for all $x \in X$ and, hence, g is a convex function; see Theorem 1.1. □

1.5 Interval analysis

In this section we summarize principles of interval analysis as given, for example, in the textbooks by Ratschek and Rokne (1988), Neumaier (1990), or Hansen (1992). Note that we assume compact intervals throughout this section.

Definition 1.9. A (compact) interval X is denoted by

$$X \;=\; [a,b] \;\subset\; \mathbb{R}$$

with $a \le b$. Moreover, the left and right endpoints are denoted by $X^L = a$ and $X^R = b$, respectively. If $X^L = X^R = z$ we sometimes use the short form $X = z = [z,z]$; that is $[z,z]$ is equivalent to z.

Next, arithmetic operations between intervals are defined as follows.

Definition 1.10. Let $X = [a,b]$ and $Y = [c,d]$ be two intervals. Then the ***interval arithmetic*** is given by

$$X \star Y \;:=\; \{x \star y \;:\; x \in X,\; y \in Y\},$$

where $\star$ denotes the addition, multiplication, subtraction, division, minimum, or maximum as long as $x \star y$ is defined for all $y \in Y$.

Due to the intermediate value theorem, $X \star Y$ again yields an interval that contains $x \star y$ for all $x \in X$ and $y \in Y$. We directly obtain the following results; see Hansen (1992).

Corollary 1.1. *Let $X = [a,b]$ and $Y = [c,d]$ be two intervals. Then*

$$\begin{aligned} X+Y &= [a+c,\, b+d], \\ X-Y &= [a-d,\, b-c], \\ X \cdot Y &= [\min\{ac,ad,bc,bd\},\, \max\{ac,ad,bc,bd\}], \\ X/Y &= [a,b] \cdot [1/d, 1/c] \text{ if } c>0, \\ \min\{X,Y\} &= [\min\{a,c\},\, \min\{b,d\}], \\ \max\{X,Y\} &= [\max\{a,c\},\, \max\{b,d\}]. \end{aligned}$$

Apart from interval arithmetics interval operations are also defined as follows.

Definition 1.11. Let $X = [a,b]$ be an interval. Then the ***interval operation*** is given by

$$op(X) := \{op(x) : x \in X\} = \left[\min_{x \in X} op(x),\, \max_{x \in X} op(x)\right],$$

where $op : X \to \mathbb{R}$ denotes a continuous function such that $op(X)$ is an interval.

In the following we assume operations such that the interval $op(X)$ can be computed easily. For example, if op is an increasing function we obtain

$$op(X) = [op(a), op(b)]$$

and if op is a decreasing function we obtain

$$op(X) = [op(b), op(a)]$$

for all intervals $X = [a,b]$. Some more examples are as follows.

Corollary 1.2. *Let $X = [a,b]$ be an interval and $n \in \mathbb{N}$. Then*

$$|X| = \begin{cases} [a,\, b] & \text{if } a \geq 0 \\ [-b,\, -a] & \text{if } b \leq 0 \\ [0,\, \max\{|a|,|b|\}] & \text{if } a \leq 0 \leq b \end{cases},$$

$$X^n = \begin{cases} [1,\, 1] & \text{if } n=0 \\ [b^n,\, a^n] & \text{if } b \leq 0 \text{ and } n \text{ even} \\ [0,\, (\max\{|a|,|b|\})^n] & \text{if } a \leq 0 \leq b \text{ and } n \text{ even} \\ [a^n,\, b^n] & \text{else} \end{cases}.$$

We remark again that interval operations can be defined for any continuous function $op : X \to \mathbb{R}$. But for our further calculations it is quite important that $op(Y)$ can be computed easily for all intervals $Y \subset X$.

Definition 1.12. An ***interval function*** $F(X_1,\ldots,X_n)$ is an interval-valued function with n intervals as arguments using interval arithmetics and interval operations as defined before.

Example 1.3. An interval function $F(X,Y)$ with two intervals X and Y as arguments is

$$F(X,Y) = \exp\left(\frac{X+Y}{Y^2+[1,1]}\right) = \exp\left(\frac{X+Y}{Y^2+1}\right).$$

For example, we obtain

$$F([0,2],[-1,1]) = \exp\left(\frac{[-1,3]}{[0,1]+[1,1]}\right) = \exp([-1,3]) = [\exp(-1),\exp(3)].$$

The following property is called ***inclusion monotonicity*** and is important for bounding operations derived from interval analysis which is presented in Chapter 3.

Lemma 1.12. *Consider an interval function $F(X_1,\ldots,X_n)$. Then*

$$F(Y_1,\ldots,Y_n) \subset F(Y_1',\ldots,Y_n')$$

for all intervals $Y_1,Y_1',\ldots,Y_n,Y_n'$ with $Y_k \subset Y_k'$ for $k=1,\ldots,n$.

Proof. See, for instance, Hansen (1992). □

In order to solve global optimization problems later on, we need the natural interval extension.

Definition 1.13. Let $f(x_1,\ldots,x_n)$ be a fixed representation (see Example 1.6) of a real-valued function with n real numbers as argument using arithmetics and operations such that the corresponding interval arithmetics and interval operations are defined.

Then the ***natural interval extension*** of $f(x_1,\ldots,x_n)$ is given by the interval function $F(X_1,\ldots,X_n)$, where arithmetics and operations are replaced by their corresponding interval arithmetics and interval operations.

Example 1.4. The natural interval extension of

$$f(x,y) = 4\cdot x^2 + \frac{\sin(y)}{x^2+1}$$

is given by

$$F(X,Y) = [4,4]\cdot X^2 + \frac{\sin(Y)}{X^2+[1,1]} = 4\cdot X^2 + \frac{\sin(Y)}{X^2+1},$$

where X and Y are intervals.

The natural interval extension leads to general lower bounds as required throughout branch-and-bound algorithms. To this end, we need the following statements which can be found, for instance, in Hansen (1992).

Lemma 1.13. *Let $F(X_1,\dots,X_n)$ be the natural interval extension of $f(x_1,\dots,x_n)$. Then*

$$f(x_1,\dots,x_n) \in F([x_1,x_1],\dots,[x_n,x_n]) = [f(x_1,\dots,x_n), f(x_1,\dots,x_n)]$$

for all $(x_1,\dots,x_n) \in X_1 \times \cdots \times X_n$.

Proof. The statement is trivial. □

Theorem 1.3 (Fundamental theorem). *Let $F(X_1,\dots,X_n)$ be the natural interval extension of $f(x_1,\dots,x_n)$. Then*

$$f(Y_1,\dots,Y_n) \subset F(Y_1,\dots,Y_n)$$

for all intervals $Y_k \subset X_k$ for $k=1,\dots,n$, where

$$f(Y_1,\dots,Y_n) = \{f(x_1,\dots,x_n) : x_k \in Y_k \text{ for } k=1,\dots,n\}.$$

Proof. Although the proof can be found, for example, in Hansen (1992), it is presented here because the result is of fundamental importance for the interval bounding operations in Chapter 3.

For any $x_k \in Y_k$ for $k=1,\dots,n$ we have $[x_k,x_k] \subset Y_k$ and therefore

$$F([x_1,x_1],\dots,[x_n,x_n]) \subset F(Y_1,\dots,Y_n);$$

see Lemma 1.12. Moreover, Lemma 1.13 yields

$$f(x_1,\dots,x_n) \in F([x_1,x_1],\dots,[x_n,x_n]) \subset F(Y_1,\dots,Y_n),$$

which proves the theorem. □

The following examples show some properties of natural interval extensions that we always should keep in mind.

Example 1.5. In general we have

$$f(X_1,\dots,X_n) \neq F(X_1,\dots,X_n).$$

For example, consider $f(x) = x^2 - 2x$ with natural interval extension

$$F(X) = X^2 - 2 \cdot X$$

and let $Y = [1,2]$. Because f is monotone increasing on Y, we find

$$f(Y) = [f(1), f(2)] = [-1,0].$$

But the natural interval extension yields

$$F(Y) = F([1,2]) = [1,4] - [2,4] = [-3,2].$$

Example 1.6. Defining

$$f_1(x) \;=\; 4(x^2-x) \;\text{ and }\; f_2(x) \;=\; (2x-1)^2-1,$$

we have $f_1(x) = f_2(x)$ for all $x \in \mathbb{R}$. This is not true for the natural interval extension and intervals. Consider the natural interval extensions

$$F_1(X) \;=\; 4\cdot(X^2-X) \;\text{ and }\; F_2(X) \;=\; (2\cdot X-1)^2-1.$$

Then, for $Y = [0,2]$, we obtain

$$F_1([0,2]) \;=\; [-8,16] \;\text{ and }\; F_2([0,2]) \;=\; [-1,8].$$

Therefore, we always assume a fixed representation of $f(x)$ if we consider the corresponding natural interval extension $F(X)$; see Definition 1.13.

Moreover, note that $F_2([0,2]) = f_2([0,2])$ because $f_2(\frac{1}{2}) = -1$ and $f_2(2) = 8$. Thus, the interval function $F_2(X)$ yields stronger bounds as we see in the remainder of this subsection.

In order to avoid these problems, we introduce the following definition which is of fundamental importance again in Chapter 3.

Definition 1.14. A fixed representation of a continuous real-valued function

$$f(x_1,\ldots,x_n)$$

is called a ***single-use expression*** if in its representation every variable x_1 to x_n occurs only once.

Example 1.7. Consider

$$f_1(x) \;=\; \exp(4(x^2-x)) \;\text{ and }\; f_2(x) \;=\; \exp((2x-1)^2-1).$$

Then f_2 is a single-use expression whereas f_1 is not, although $f_1(x) = f_2(x)$ for all $x \in \mathbb{R}$. As a second example, consider

$$f_1(x,y) \;=\; (x+y)^2+3 \;\text{ and }\; f_2(x,y) \;=\; x^2+2xy+3+y^2.$$

Here, only f_1 is a single-use expression, although again $f_1(x,y) = f_2(x,y)$ for all $(x,y) \in \mathbb{R}^2$.

The reason for the definition of single-use expressions is the following result.

Theorem 1.4. *Let $F(X_1,\ldots,X_n)$ be the natural interval extension of a single-use expression $f(x_1,\ldots,x_n)$. Then*

$$f(Y_1,\ldots,Y_n) \;=\; F(Y_1,\ldots,Y_n) \;=\; \left[\min_{\substack{x_k\in Y_k\\ k=1,\ldots,n}} f(x_1,\ldots,x_n),\; \max_{\substack{x_k\in Y_k\\ k=1,\ldots,n}} f(x_1,\ldots,x_n)\right]$$

for all intervals $Y_k \subset X_k$ *for* $k = 1, \ldots, n$.

Proof. See, for instance, Neumaier (1990). □

Remark 1.1. Let $f(x_1, \ldots, x_n)$ be a single-use expression, let $F(X_1, \ldots, X_n)$ be the corresponding natural interval extension, and consider some intervals $Y_k \subset X_k$ for $k = 1, \ldots, n$. Then, from Theorem 1.4, we know that there exists a $(y_1, \ldots, y_n) \in Y_1 \times \ldots \times Y_n$ such that

$$f(y_1, \ldots, y_n) = F(Y_1, \ldots, Y_n)^L.$$

But note that it is in general not an easy task to find such a $(y_1, \ldots, y_n)$ inasmuch as we need to solve the problem

$$\min_{\substack{x_k \in Y_k \\ k=1,\ldots,n}} f(x_1, \ldots, x_n).$$

For the case of general interval functions, we need the following definition; see Hansen (1992).

Definition 1.15. An interval function $F(X_1, \ldots, X_n)$ is called an ***interval inclusion function*** of $f(x_1, \ldots, x_n)$ if

$$f(Y_1, \ldots, Y_n) \subset F(Y_1, \ldots, Y_n)$$

for all intervals $Y_k \subset X_k$ for $k = 1, \ldots, n$.

Hence, every interval inclusion function leads to lower bounds on the objective function as required throughout the geometric branch-and-bound algorithm suggested in the following chapter. As an example, the natural interval extension yields an interval inclusion function due to Theorem 1.3.

Chapter 2
The geometric branch-and-bound algorithm

Abstract The aim of this chapter is the presentation of the fundamental geometric branch-and-bound algorithm including a general convergence theory. To this end, we start with a literature review of these approaches and their applications to facility location problems in Section 2.1. Next, we give some notations and formally define bounding operations in Section 2.2 before the geometric branch-and-bound prototype algorithm is presented and discussed in Section 2.3. The main contribution of the present chapter can be found in Sections 2.4 and 2.5. Therein, we define the rate of convergence and results about the termination of the algorithm are given. Note that parts of these sections have been published in Schöbel and Scholz (2010b).

2.1 Literature review

Over the last decades, geometric branch-and-bound methods became more and more important solution algorithms for global optimization problems. Although all these methods are based on the same ideas, they differ in particular in the way of calculating lower bounds.

2.1.1 General branch-and-bound algorithms

Some general branch-and-bound techniques for continuous problems can be found in Horst and Tuy (1996) and Horst et al. (2000). Using Lipschitzian functions, lower bounds can be constructed as suggested in Hansen and Jaumard (1995). In Horst and Thoai (1999) and Tuy and Horst (1988), branch-and-bound techniques for d.c. functions are considered. Global optimization using interval analysis is discussed in Ratschek and Rokne (1988), Hansen (1992), and Ratschek and Voller (1991). Finally, an extension for multicriteria problems was first introduced in Ichida and Fujii (1990).

2.1.2 Branch-and-bound methods in location theory

Moreover, some particular algorithms for facility location problems can be found in the literature. One of the first geometric branch-and-bound approaches in the area of facility location problems was suggested by Hansen et al. (1985): the ***big square small square*** technique for some facility location problems on the plane. Plastria (1992) generalized this method to the ***generalized big square small square*** technique. Using triangles instead of squares, Drezner and Suzuki (2004) proposed the ***big triangle small triangle*** method. General lower bounds derived from d.c. programming were suggested in Drezner (2007) and have been improved in Blanquero and Carrizosa (2009). All these techniques are branch-and-bound solution methods for problems with two variables, thus Schöbel and Scholz (2010a) suggested the ***big cube small cube*** technique for facility location problems with multiple variables. Tuy et al. (1995) and Tuy (1996) discussed a d.c. decomposition for general facility location problems that also leads to lower bounds. Extensions to multicriteria location problems can be found in Fernández and Tóth (2007, 2009) and Scholz (2010).

2.1.3 Applications to special facility location problems

All these methods have been used as solution algorithms for a wide range of facility location problems. Several examples of planar problems solved by the big triangle small triangle method can be found in Drezner and Suzuki (2004) and Drezner (2007). In Romero-Morales et al. (1997), semiobnoxious models were solved using the big square small square technique. Tóth et al. (2009), Drezner and Drezner (2004), Fernández et al. (2007b), and Bello et al. (2010) discussed Huff-like competitive problems. In Schöbel and Scholz (2010a), the multisource Weber problem and the median circle problem were considered. Moreover, Drezner and Nickel (2009a,b) solved the general ordered median problem on the plane. Equity models were studied in Drezner and Drezner (2007) and some further competition location models were solved in Fernández et al. (2006, 2007a) using lower bounds derived from interval analysis. In Zaferanieh et al. (2008), the authors considered a problem with different distance measures in different regions. Locating objects on the plane were studied in Blanquero et al. (2009) and solved using d.c. decompositions. Integrated scheduling and location problems have been solved in Kalsch and Drezner (2010). Finally, geometric branch-and-bound methods in multicriteria facility location problems can be found in Skriver and Anderson (2003), Fernández and Tóth (2009), and Scholz (2010).

2.2 Notations

Before we present the geometric branch-and-bound prototype algorithm, we introduce several notations that are needed in the following chapters.

Definition 2.1. A compact ***box*** or ***hyperrectangle*** with sides parallel to the axes is denoted by

$$X = [x_1^L, x_1^R] \times \cdots \times [x_n^L, x_n^R] \subset \mathbb{R}^n.$$

The ***diameter*** of a box $X \subset \mathbb{R}^n$ is

$$\delta(X) = \max\{\|x-\tilde{x}\|_2 : x, \tilde{x} \in X\} = \sqrt{(x_1^R - x_1^L)^2 + \cdots + (x_n^R - x_n^L)^2}$$

and the ***center*** of a box $X \subset \mathbb{R}^n$ is defined by

$$c(X) = \left(\frac{1}{2}(x_1^L + x_1^R), \ldots, \frac{1}{2}(x_n^L + x_n^R)\right).$$

Next, we define bounding opterations as follows.

Definition 2.2. Let $X \subset \mathbb{R}^n$ be a box and consider $f : X \to \mathbb{R}$. A ***bounding operation*** is a procedure to calculate for any subbox $Y \subset X$ a ***lower bound*** $LB(Y) \in \mathbb{R}$ with

$$LB(Y) \leq f(x) \text{ for all } x \in Y$$

and to specify a point $r(Y) \in Y$. Formally, we obtain the bounding operation

$$(LB(Y), r(Y))$$

for all subboxes $Y \subset X$.

Several general bounding operations are derived in Chapter 3. Moreover, note that the choice of $r(Y)$ is important for our theoretical results because it affects the theoretical rate of convergence; see Example 2.1.

2.3 The geometric branch-and-bound algorithm

We consider the minimization of a continuous function

$$f : X \to \mathbb{R},$$

where we assume a box $X \subset \mathbb{R}^n$ as the feasible area;

$$X = [x_1^L, x_1^R] \times \cdots \times [x_n^L, x_n^R] \subset \mathbb{R}^n.$$

The general idea of all the geometric branch-and-bound algorithms cited in Section 2.1 is the same: subboxes of the feasible area are bounded from below making use of a bounding operation as defined before. If the bounds are not sharp enough, some boxes are split into smaller ones according to a given ***splitting rule***; see Subsection 2.3.2. This procedure repeats until the algorithm finds an optimal solution $x^* \in X$ within an absolute accuracy of $\varepsilon > 0$.

To sum up, for the following geometric branch-and-bound algorithm assume an objective function f and a feasible box X. Moreover, we need a bounding operation, a splitting rule, and an absolute accuracy of $\varepsilon > 0$.

1. Let $\mathscr{X}$ be a list of boxes and initialize $\mathscr{X} := \{X\}$.
2. Apply the bounding operation to X and set $UB := f(r(X))$ and $x^* := r(X)$.
3. If $\mathscr{X} = \emptyset$, the algorithm stops. Else set

$$\delta_{\max} := \max\{\delta(Y) : Y \in \mathscr{X}\}.$$

4. Select a box $Y \in \mathscr{X}$ with $\delta(Y) = \delta_{\max}$ and split it according to the splitting rule into s congruent smaller subboxes Y_1 to Y_s.
5. Set $\mathscr{X} = (\mathscr{X} \setminus Y) \cup \{Y_1, \ldots, Y_s\}$; that is delete Y from $\mathscr{X}$ and add $Y_1, \ldots, Y_s$.
6. Apply the bounding operation to $Y_1, \ldots, Y_s$ and set

$$UB = \min\{UB, f(r(Y_1)), \ldots, f(r(Y_s))\}.$$

   If $UB = f(r(Y_k))$ for a $k \in \{1, \ldots, s\}$, set $x^* = r(Y_k)$.
7. For all $Z \in \mathscr{X}$, if $LB(Z) + \varepsilon \geq UB$ set $\mathscr{X} = \mathscr{X} \setminus Z$. If UB has not changed it is sufficient to check only the subboxes Y_1 to Y_s.
8. Whenever possible, apply some further discarding tests; see Subsection 2.3.4.
9. Return to Step 3.

Recall that general bounding operations are collected in Chapter 3. In the following, we discuss the most important steps of the algorithm.

2.3.1 Selection rule and accuracy

In the proposed algorithm, a box with the largest diameter is selected for a split into some subboxes; see Step 3. Another selection rule that can be found in the literature is as follows. Select a box $Y \in \mathscr{X}$ with smallest lower bound; that is select $Y \in \mathscr{X}$ such that

$$LB(Y) = LB_{\min} := \min\{LB(Z) : Z \in \mathscr{X}\}.$$

Moreover, note that ε is an absolute accuracy and not a relative accuracy as often used in the literature. Our proposed selection rule and absolute accuracy are necessary for the theoretical results given in the next section.

2.3.2 Splitting rule

The selected box Y in Step 4 has to be split into s subboxes Y_1 to Y_s such that

$$Y = \bigcup_{k=1}^{s} Y_k.$$

For boxes in small dimensions, say $n \leq 3$, we suggest a split into $s = 2^n$ congruent subboxes. In higher dimensions, boxes can be bisected perpendicular to the direction of the maximum width component in two subboxes. Some more sophisticated splitting rules and numerical results about them can be found in Casado et al. (2000), Csallner et al. (2000), and Markót et al. (1999).

2.3.3 Shape of the sets

Although some more general shapes of the sets Y are also possible, for example, triangles if $n = 2$ as proposed in Drezner and Suzuki (2004), in this work we restrict ourselves to boxes. However, most of the bounding operations presented in Chapter 3 are also suitable for polyhedra instead of boxes.

Moreover, note that the feasible area X can also be given as a union of a finite number of boxes or polyhedra if a more general shape of the feasible area is required.

2.3.4 Discarding tests

If it can be shown that a box Y does not contain a global minimum, Y can be discarded from further consideration. Rules to do so are called ***discarding tests***; see Step 8 of the algorithm. For example, assume that the objective function f is differentiable. If it can be shown that $0 \notin \{\nabla f(x) \ : \ x \in Y\}$, the box Y can be discarded. Using tools from interval analysis, some general discarding tests are collected in Tóth et al. (2009) and Fernández et al. (2007b), among others.

2.4 Rate of convergence

Before we present a general convergence theory, let us start with the concept of consistent bounding operations.

Definition 2.3. A bounding operation is called ***consistent*** if

$$\lim_{k\to\infty}\Big(f(r(Y_k)) - LB(Y_k)\Big) = 0$$

holds for all sequences $(Y_k)_{k\in\mathbb{N}}$ of boxes Y_k with $\lim_{k\to\infty}\delta(Y_k) = 0$.

The most important definition for our further consideration is the following one.

Definition 2.4. Let $X \subset \mathbb{R}^n$ be a box and $f : X \to \mathbb{R}$. Furthermore, consider the minimization problem

$$\min_{x\in X} f(x).$$

We say a bounding operation has the ***rate of convergence*** $p \in \mathbb{N}$ if there exists a fixed constant $C > 0$ such that

$$f(r(Y)) - LB(Y) \leq C\cdot\delta(Y)^p \tag{2.1}$$

for all boxes $Y \subset X$.

We remark that the rate of convergence does not depend on the norm used for the diameter $\delta(Y)$ of a box Y because all norms are equivalent in $\mathbb{R}^n$.

A related concept concerning this topic can be found in Tuy and Horst (1988) and Horst and Tuy (1996). Therein, general convergence criteria without the explicit use of bounding operations are given. In Plastria (1992), a convergence theorem based on Lipschitzian optimization is given for planar facility location problems discussed therein. Drezner (2007) treated location problems where the objective function is a sum of d.c. functions of the distance, and the quality of the proposed bounds is discussed.

In the following we introduce a convergence theory using the new concept of the rate of convergence suggested in Definition 2.4. A related concept for the specific use of interval analysis methods was introduced in Csallner and Csendes (1996). Their definition of the rate of convergence does not require the specification of a point $r(Y)$ for all boxes Y but they used the exact range $f(Y)$ of f on Y which cannot be derived in general. Their concept was only analyzed from the empirical point of view; see Tóth and Csendes (2005).

Lemma 2.1. *Consider a bounding operation that has a rate of convergence of $p \geq 1$. Then the bounding operation is consistent.*

Proof. Consider an arbitrary sequence $(Y_k)_{k\in\mathbb{N}}$ of boxes Y_k with

$$\lim_{k\to\infty}\delta(Y_k) = 0.$$

The bounding operation has a rate of convergence of $p \geq 1$, therefore it follows that

$$f(r(Y_k)) - LB(Y_k) \leq C\cdot\delta(Y_k)^p$$

for all $k \in \mathbb{N}$. Hence,

$$\lim_{k\to\infty}\Big(f(r(Y_k)) - LB(Y_k)\Big) \;\leq\; \lim_{k\to\infty} C\cdot\delta(Y_k)^p \;=\; 0.$$

Furthermore, with $LB(Y_k) \leq f(x)$ for all $k \in \mathbb{N}$ and all $x \in Y_k$, we have

$$\lim_{k\to\infty}\Big(f(r(Y_k)) - LB(Y_k)\Big) \;\geq\; 0.$$

Thus, we obtain

$$\lim_{k\to\infty}\Big(f(r(Y_k)) - LB(Y_k)\Big) \;=\; 0,$$

which proves the lemma. □

The following example shows that the choice of $r(Y)$ can have an impact on the rate of convergence.

Example 2.1. Consider the linear function $f(x) = \alpha x + \beta$ with $\alpha, \beta \in \mathbb{R}$ and for all boxes $Y = [y^L, y^R] \subset \mathbb{R}$ consider the bounding operation

$$LB_1(Y) \;=\; \min\{f(y^L), f(y^R)\} \;\text{ and }\; r_1(Y) \;=\; c(Y).$$

We then easily find

$$f(r_1(Y)) - LB_1(Y) \;=\; \frac{|\alpha|}{2}\cdot(y^R - y^L) \;=\; \frac{|\alpha|}{2}\cdot\delta(Y).$$

Moreover, the bounding operation

$$LB_2(Y) \;=\; LB_1(Y) \;\text{ and }\; r_2(Y) \;=\; \arg\min_{x\in\{y^L, y^R\}} f(x).$$

yields

$$f(r_2(Y)) - LB_2(Y) \;=\; 0.$$

Both bounding operations are consistent, but the first one has a rate of convergence of not larger than $p = 1$ and the second one $p = \infty$.

Before we consider the convergence of the algorithm, the next lemma shows that bounding operations are additive.

Lemma 2.2. *Let $X \subset \mathbb{R}^n$ be a box and consider $f_1, f_2 : X \to \mathbb{R}$. Furthermore, assume that for f_1 a bounding operation $(LB_1(Y), r(Y))$ and for f_2 a bounding operation $(LB_2(Y), r(Y))$ in each case with a rate of convergence of p is known. Then*

$$(LB_1(Y) + LB_2(Y), r(Y))$$

is a bounding operation for $f_1 + f_2$ that has a rate of convergence of p.

Proof. Consider an arbitrary box $Y \subset X$ and define $g(x) := f_1(x) + f_2(x)$. Because

$$LB_1(Y) \;\leq\; f_1(x) \;\text{ and }\; LB_2(Y) \;\leq\; f_2(x) \;\text{ for all } x \in Y,$$

we directly obtain

$$LB_1(Y) + LB_2(Y) \leq f_1(x) + f_2(x) = g(x) \text{ for all } x \in Y.$$

Hence, $LB(Y) := LB_1(Y) + LB_2(Y)$ is a lower bound for g. Furthermore, we know that

$$f_1(r(Y)) - LB_1(Y) \leq C_1 \cdot \delta(Y)^p,$$
$$f_2(r(Y)) - LB_2(Y) \leq C_2 \cdot \delta(Y)^p$$

for all boxes $Y \subset X$ which directly yields

$$\begin{aligned} g(r(Y)) - LB(Y) &= (f_1(r(Y)) + f_2(r(Y))) - (LB_1(Y) + LB_2(Y)) \\ &= (f_1(r(Y)) - LB_1(Y)) + (f_2(r(Y)) - LB_2(Y)) \\ &\leq C_1 \cdot \delta(Y)^p + C_2 \cdot \delta(Y)^p = (C_1 + C_2) \cdot \delta(Y)^p. \end{aligned}$$

Thus, the rate of convergence of p for the bounding operation

$$(LB_1(Y) + LB_2(Y),\ r(Y))$$

is shown. □

2.5 Convergence theory

In this section we restrict ourselves to the case where we split all selected boxes into $s = 2^n$ subboxes, although similar results can be found for other splitting rules.

Theorem 2.1. *Consider the geometric branch-and-bound algorithm with a bounding operation that has a rate of convergence of $p \geq 1$ and a constant of $C > 0$. Furthermore, assume that each selected box throughout the algorithm is split into $s = 2^n$ congruent smaller boxes and that the algorithm stops with an absolute accuracy of $\varepsilon > 0$.*

Then the worst case number of iterations for the geometric branch-and-bound algorithm can be bounded by

$$\sum_{k=0}^{z} 2^{n \cdot k} \text{ where } z = \left\lceil \log_2 \frac{\delta(X)}{\left(\frac{\varepsilon}{C}\right)^{1/p}} \right\rceil \tag{2.2}$$

and $\delta(X)$ is the diameter of the feasible box X.

Proof. Denote the list of boxes throughout the geometric branch-and-bound algorithm by $\mathscr{X}$ and note that

$$UB - LB(Y) \leq f(r(Y)) - LB(Y) \leq C \cdot \delta(Y)^p$$

for all $Y \in \mathscr{X}$ in every iteration of the algorithm. If $\delta(Y)$ is small enough for all boxes $Y \in \mathscr{X}$, we have

$$UB - LB(Y) \leq C \cdot \delta(Y)^p \leq \varepsilon$$

for all $Y \in \mathscr{X}$ because $C > 0$ is a fixed constant. This is a sufficient condition for the termination of the algorithm because the termination rule is satisfied for all $Y \in \mathscr{X}$.

Next, we calculate how small $\delta(Y)$ needs to be in the worst case. From $C \cdot \delta(Y)^p \leq \varepsilon$ one has

$$\delta(Y) \leq \left(\frac{\varepsilon}{C}\right)^{1/p} =: K > 0.$$

As a connection between $\delta(X)$ and $\delta(Y)$, consider the smallest value of $N \in \mathbb{N}$ such that

$$\frac{\delta(X)}{N} \leq \delta(Y) \leq K; \text{ that is } \frac{\delta(X)}{K} \leq N.$$

Thus, the length of each of the n sides of X should be smaller than or equal to the original length divided by N. Hence, if the initial box X is split into N^n congruent smaller boxes, the algorithm terminates.

We now want to count the number of iterations needed. After the first iteration, we split X into 2^n smaller boxes. If $N \leq 2$, the algorithm stops. Otherwise we need 2^n more iterations to split in the worst case all smaller boxes. Again, if $N > 4$, we need $2^n \cdot 2^n$ additional splits and so on. For the general case, define

$$z = \min\{k : k \in \mathbb{N}, N \leq 2^k\} = \lceil \log_2 N \rceil .$$

Thus, the algorithm stops after

$$1 + 2^n + (2^n)^2 + \cdots + (2^n)^z = \sum_{k=0}^{z} 2^{n \cdot k}$$

iterations. This is exactly the same as stated in the theorem. □

Corollary 2.1. *Consider the geometric branch-and-bound algorithm with a bounding operation that has a rate of convergence of $p \geq 1$. Then for any $\varepsilon > 0$ the algorithm stops after finitely many steps with an ε-optimal solution.*

Proof. Follows directly from Theorem 2.1. □

The following examples show the importance of the rate of convergence.

Example 2.2. Assume an initial box X with $\delta(X) = 1$ and an absolute accuracy of $\varepsilon = 10^{-2}$. Moreover, assume the constant $C = 10$ for the rate of convergence. Table 2.1 summarizes the worst case numbers of iterations derived from Equation (2.2) for various values of the rate of convergence p and the dimension n.

Thus, it is important to use a bounding operation with a greatest possible rate of convergence.

n	p	# (Iterations)
2	1	1,398,101
2	2	1,365
2	3	341
3	1	1,227,133,513
3	2	37,449
3	3	4,681

Table 2.1 Worst case number of iterations for various values of the rate of convergence p and the dimension n.

Example 2.3. Even if we suppose that every mth box can be discarded throughout the branch-and-bound phase, the worst case number of iterations can be bounded by

$$1+\left(2^n-\frac{2^n}{m}\right)+\left(2^n-\frac{2^n}{m}\right)^2+\cdots+\left(2^n-\frac{2^n}{m}\right)^z.$$

Using again an initial box X with $\delta(X)=1$, an absolute accuracy of $\varepsilon=10^{-2}$, and $C=10$, Table 2.2 shows the worst case runtimes for the case $n=3$.

n	p	m	# (Iterations)
3	1	4	72,559,411
3	2	4	9,331
3	3	4	1,555
3	1	2	1,398,101
3	2	2	1,365
3	3	2	341

Table 2.2 Worst case number of iterations for various values of the rate of convergence p if every mth box is discarded.

Chapter 3
Bounding operations

Abstract In the previous chapter, the geometric branch-and-bound prototype algorithm was formulated and discussed. However, the most important choice for fast branch-and-bound algorithms is good bounding operations; see Example 2.2. Therefore, several bounding operations are derived in this chapter; see Sections 3.1 to 3.9, and for all bounding operations we prove the theoretical rate of convergence as defined in Section 2.4. Apart from our theoretical results, all findings are justified by numerical experiments in Section 3.10. We remark that almost all results collected in this chapter can be found in Schöbel and Scholz (2010b) and Scholz (2011b).

3.1 Concave bounding operation

Here and in the following sections we again consider optimization problems

$$\min_{x \in X} f(x)$$

and our goal is to derive bounding operations; that is for all boxes $Y \subset X$ we want to find a lower bound $LB(Y)$ and specify a point $r(Y) \in Y$; see Definition 2.2.

A very simple bounding operation is the following one. Assume that $f : X \to \mathbb{R}$ is concave on X and denote the set of the 2^n vertices of a box Y by $V(Y)$. Then, due to Lemma 1.4, we directly obtain the ***concave bounding operation***

$$LB(Y) := \min_{v \in V(Y)} f(v) \quad \text{and} \quad r(Y) \in \arg\min_{v \in V(Y)} f(v); \tag{3.1}$$

see Figure 3.1. Note that in the literature mostly concave minimization problems are considered; see, for instance, Horst and Tuy (1996). However, the proposed bounding operation is also valid for quasiconcave functions; see Lemma 1.4.

Theorem 3.1. *Assume that $f : X \to \mathbb{R}$ is concave on X.*

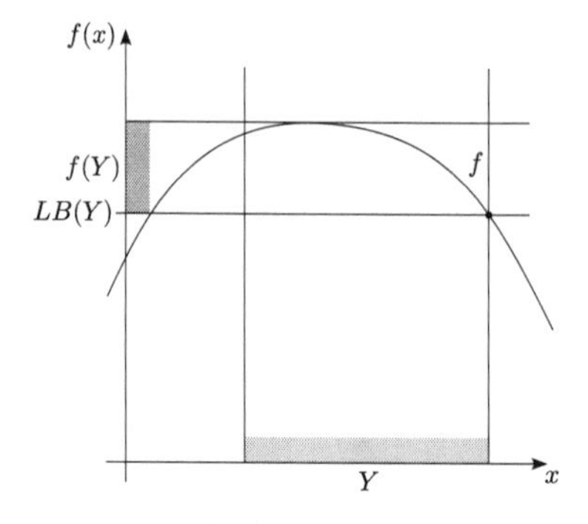

Fig. 3.1 Illustration for the concave bounding operation.

Then the concave bounding operation has a rate of convergence of $p = \infty$.

Proof. Due to Lemma 1.4, we have

$$f(r(Y)) - LB(Y) \; = \; 0 \; \leq \; C \cdot \delta(Y)^p$$

for all boxes $Y \subset X$ and for any $C > 0$ and $p \in \mathbb{N}$. Hence, the concave bounding operation has the rate of convergence of $p = \infty$. □

The following result clarifies again that the choice of $r(Y)$ is important for the theoretical rate of convergence. To this end, consider the bounding operation

$$LB(Y) \; := \; \min_{v \in V(Y)} f(v) \quad \text{and} \quad r(Y) \; := \; c(Y), \tag{3.2}$$

where $c(Y)$ is the center of Y; see Definition 2.1. For concave and Lipschitzian functions we then obtain the following result.

Theorem 3.2. *Assume that $f : X \to \mathbb{R}$ is a concave and Lipschitzian function with constant L on X.*

Then the bounding operation as defined in Equation (3.2) *has a rate of convergence of $p = 1$.*

Proof. Let $u \in \arg\min\{f(v) \; : \; v \in V(Y)\} \subset Y$ and define $c = c(Y)$. Then we find

$$f(r(Y)) - LB(Y) \; = \; f(c) - f(u) \; \leq \; L \cdot \|c - u\|_2 \; \leq \; \frac{L}{2} \cdot \delta(Y)$$

for all boxes $Y \subset X$. □

Finally, the following result shows that we cannot expect more.

Example 3.1. Let $f(x) = -x^2$ and assume the initial box $X = [0, 2]$. Using the sequence $Y_\mu = [1 - \mu, 1 + \mu] \subset X$ with $0 < \mu < 1$, we obtain

$$\frac{f(r(Y_\mu)) - LB(Y_\mu)}{\delta(Y_\mu)^2} = \frac{-1+(1+\mu)^2}{4\mu} = \frac{1}{4} + \frac{1}{2\mu}.$$

Inasmuch as this expression is unbounded for $\mu \to 0$, we cannot find a fixed constant $C > 0$ such that

$$f(r(Y_\mu)) - LB_1(Y_\mu^2) \le C \cdot \delta(Y_\mu)^2$$

for all $\mu > 0$. Hence, the bounding operation as defined in Equation (3.2) has a rate of convergence not higher than $p = 1$.

Remark 3.1. The concave bounding operation is very simple and we can easily solve the problem up to an optimal solution for all boxes Y. However, this bounding operation can only be used if we consider a minimization problem with a concave or quasiconcave objective function as mentioned before. Moreover, we demonstrated that the theoretical rate of convergence strongly depends on the choice of $r(Y)$.

3.2 Lipschitzian bounding operation

The next bounding operation is derived from Lipschitzian optimization; see Hansen and Jaumard (1995). Assume that $f : X \to \mathbb{R}$ is a Lipschitzian function on X with Lipschitzian constant $L > 0$ and further assume that an upper bound $A \ge L$ is known.

With $f(x) - f(y) \le |f(x) - f(y)|$, we obtain for $c = c(Y) \in Y$

$$\begin{aligned} f(c) &\le f(x) + |f(c) - f(x)| \le f(x) + L \cdot \|c - x\|_2 \\ &\le f(x) + A \cdot \|c - x\|_2 \le f(x) + \frac{1}{2} A \cdot \delta(Y) \end{aligned}$$

for all $x \in Y$; see Figure 3.2.

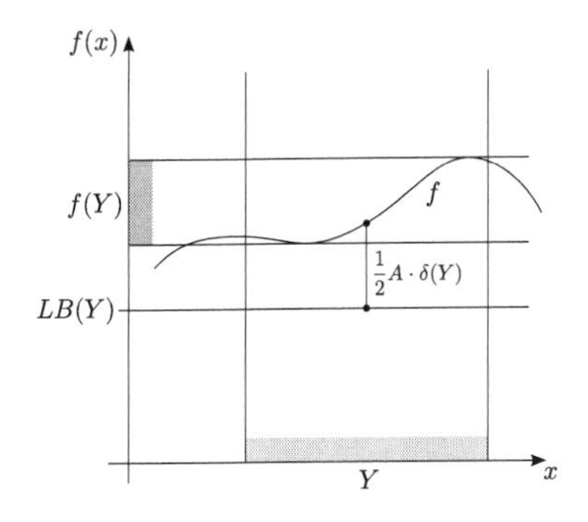

Fig. 3.2 Illustration for the Lipschitzian bounding operation.

Therefore, we have the ***Lipschitzian bounding operation***

$$LB(Y) := f(c(Y)) - \frac{1}{2}A \cdot \delta(Y) \quad \text{and} \quad r(Y) := c(Y). \tag{3.3}$$

Again, we can easily derive the rate of convergence.

Theorem 3.3. *Assume that $f : X \to \mathbb{R}$ is a Lipschitzian function on X with constant L and let $A \geq L$.*

Then the Lipschitzian bounding operation has a rate of convergence of (at least) $p = 1$.

Proof. For any box $Y \subset X$ the Lipschitzian bounding operation yields

$$LB(Y) = f(r(Y)) - \frac{1}{2}A \cdot \delta(Y).$$

With $C = A/2$ we directly obtain

$$f(r(Y)) - LB(Y) = \frac{1}{2}A \cdot \delta(Y) = C \cdot \delta(Y)$$

which shows the rate of convergence of $p = 1$. □

The following example shows that we have exactly $p = 1$.

Example 3.2. Let $f(x) = x^2$ and assume the initial box $X = [-2,2]$. Then f is a Lipschitzian function on X with constant $L = 4$.

Using the sequence $Y_\mu = [-\mu, \mu] \subset X$ with $0 < \mu < 2$ and $A = L = 4$, we obtain

$$\frac{f(r(Y_\mu)) - LB(Y_\mu)}{\delta(Y_\mu)^2} = \frac{2 \cdot \delta(Y_\mu)}{\delta(Y_\mu)^2} = \frac{2}{\delta(Y_\mu)} = \frac{1}{\mu}.$$

Because $1/\mu$ is unbounded for $\mu \to 0$, we cannot find a fixed constant $C > 0$ such that

$$f(r(Y_\mu)) - LB_1(Y_\mu) \leq C \cdot \delta(Y_\mu)^2$$

for all $\mu > 0$. Therefore, the Lipschitzian bounding operation has a rate of convergence of exactly $p = 1$.

Remark 3.2. First of all note that the rate of convergence of $p = 1$ does not depend on the choice of $r(Y)$; that is for all $r(Y) \in Y$ we obtain the same rate of convergence because f is a Lipschitzian function. Moreover, in general it is not an easy task to find the required constant $A \geq L$. However, if such an A is known, the calculation of the lower bounds is very easy and requires less computational effort compared to the following bounding operations.

3.3 D.c. bounding operation

The d.c. bounding operation can be found, for example, in Horst and Thoai (1999) and Tuy and Horst (1988). Assume that f is a d.c. function on Y, say

$$f(x) = g(x) - h(x) \text{ for all } x \in Y,$$

where $g, h : X \to \mathbb{R}$ are convex functions, and set $c = c(Y) \in Y$. Because g is convex, there exists a subgradient ξ of g at c; see Lemma 1.5. Hence, we have

$$a(x) := g(c) + \xi^T (x - c) \leq g(x) \text{ for all } x \in Y.$$

Together we obtain

$$m(x) := a(x) - h(x) = g(c) + \xi^T (x - c) - h(x) \leq f(x) \text{ for all } x \in Y. \quad (3.4)$$

Note that m is concave because the sum of concave functions is concave again and a and $-h$ are concave; see also Figure 3.3. Moreover, denote again by $V(Y)$ the set of the 2^n vertices of Y. Then the ***d.c. bounding operation*** is given by

$$LB(Y) := \min_{v \in V(Y)} m(v) \quad \text{and} \quad r(Y) \in \arg \min_{v \in V(Y)} m(v). \quad (3.5)$$

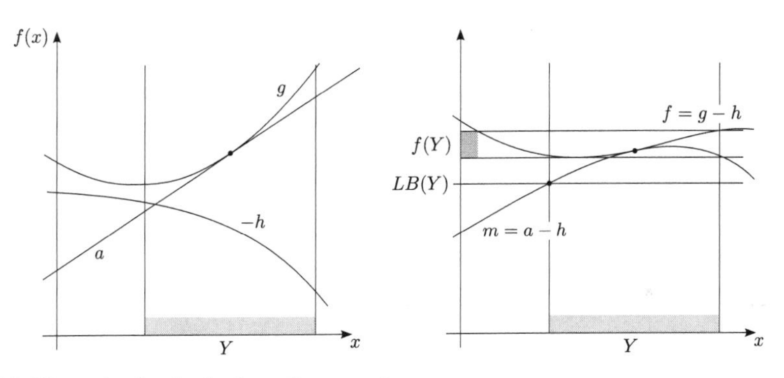

Fig. 3.3 Illustration for the d.c. bounding operation.

D.c. bounding operations have been used for several location problems; see Tuy (1996), Drezner and Suzuki (2004), Drezner and Drezner (2004), Blanquero et al. (2009), and Schöbel and Scholz (2010a).

We now discuss the rate of convergence of the d.c. bounding operation. To this end, we first of all introduce the following general result.

Lemma 3.1. *Consider a box $X \subset \mathbb{R}^n$, a subbox $Y \subset X$ with $c = c(Y)$, and some arbitrary functions $f, m : X \to \mathbb{R}$. Moreover, assume that f and m are twice continuously differentiable on X and suppose that*

$$f(c) = m(c) \text{ and } \nabla f(c) = \nabla m(c).$$

Then there exists a fixed constant $C > 0$ that does not depend on Y with

$$f(x) - m(x) \leq C \cdot \delta(Y)^2 \text{ for all } x \in Y.$$

Proof. Defining $q : X \to \mathbb{R}$ by $q(x) = f(x) - m(x)$, we have $q(c) = \nabla q(c) = 0$ and the second-order Taylor expansion of q at c yields

$$\begin{aligned} q(x) &= q(c) + \nabla q(c)^T \cdot (x-c) + \frac{1}{2} \cdot (x-c)^T \cdot D^2 q(\vartheta(x)) \cdot (x-c) \\ &= \frac{1}{2} \cdot (x-c)^T \cdot D^2 q(\vartheta(x)) \cdot (x-c), \end{aligned}$$

where $\vartheta(x) \in Y$ for all $x \in Y$ and $D^2 q(\vartheta(x))$ is the Hessian matrix of q at $\vartheta(x)$.

Next, consider $D^2 q(\vartheta)$ for an arbitrary $\vartheta \in X$. All second partial derivatives of q are continuous on X, therefore the Hessian matrix $D^2 q(\vartheta)$ is symmetric. Hence, for any $\vartheta \in X$ we obtain a decomposition of the form

$$D^2 q(\vartheta) = Q(\vartheta) \cdot \Lambda(\vartheta) \cdot Q(\vartheta)^T,$$

where $Q(\vartheta)$ is orthogonal and $\Lambda(\vartheta) = \operatorname{diag}(\lambda_1(\vartheta), \ldots, \lambda_n(\vartheta))$ is a diagonal matrix that consists of the real eigenvalues of $D^2 q(\vartheta)$. Define

$$\begin{aligned} \theta &:= \max\{|\lambda_k(\vartheta)| \;:\; k = 1, \ldots, n,\; \vartheta \in X\} \\ &= \max\{|\lambda| \;:\; \lambda \text{ is an eigenvalue of } D^2 q(\vartheta) \text{ for a } \vartheta \in X\}. \end{aligned}$$

Note that $\theta < \infty$ because $D^2 q(x)$ is continuous on X. Moreover, we define

$$y(x) := Q(\vartheta(x))^T \cdot (x-c).$$

Because $\vartheta(x) \in Y$ for all $x \in Y$ and $Q(\vartheta)$ is orthogonal for all $\vartheta \in Y$, we have

$$\|y(x)\|_2 = \|Q(\vartheta(x))^T \cdot (x-c)\|_2 = \|(x-c)\|_2.$$

Using these results, we obtain

$$\begin{aligned} f(x) - m(x) &= \frac{1}{2} \cdot (x-c)^T \cdot D^2 q(\vartheta(x)) \cdot (x-c) \\ &= \frac{1}{2} \cdot (x-c)^T \cdot Q(\vartheta(x)) \cdot \Lambda(\vartheta(x)) \cdot Q(\vartheta(x))^T \cdot (x-c) \\ &= \frac{1}{2} \cdot y(x)^T \cdot \Lambda(\vartheta(x)) \cdot y(x) \\ &\leq \frac{1}{2}\theta \cdot y(x)^T y(x) = \frac{1}{2}\theta \cdot \|y(x)\|_2^2 \\ &= \frac{1}{2}\theta \cdot \|(x-c)\|_2^2 \leq \frac{1}{2}\theta \cdot \left(\frac{\delta(Y)}{2}\right)^2 = \frac{\theta}{8} \cdot \delta(Y)^2 \end{aligned}$$

for all $x \in Y$ which proves the statement. □

Using this lemma, we easily obtain the rate of convergence for the d.c. bounding operation.

Theorem 3.4. *Assume that $f : X \to \mathbb{R}$ is a d.c. function with d.c. decomposition*

$$f(x) = g(x) - h(x)$$

and assume that g is twice continuously differentiable on X.

Then the d.c. bounding operation has a rate of convergence of $p = 2$.

Proof. With $c = c(Y)$ and $\nabla g(c) \in \mathbb{R}^n$ recall that

$$m(x) = g(c) + \nabla g(c)^T (x - c) - h(x);$$

see Equation (3.4). Hence, we have $\nabla m(x) = \nabla g(c) - \nabla h(x)$ and we get

$$m(c) = g(c) - h(c) = f(c),$$
$$\nabla m(c) = \nabla g(c) - \nabla h(c) = \nabla f(c).$$

Therefore, Lemma 3.1 yields

$$f(r(Y)) - LB(Y) = f(r(Y)) - m(r(Y)) \leq C \cdot \delta(Y)^2$$

for a $C > 0$ that does not depend on Y. This proves the rate of convergence of $p = 2$.
□

The following example shows that Theorem 3.4 cannot be improved: a rate of convergence of $p = 2$ is the highest we can expect.

Example 3.3. Consider $f(x) = x^3$ and the initial box $X = [0, 2]$. The function f is convex on X and therefore a d.c. function. For the d.c. bounding operation we hence set $g = f$ and $h = 0$.

For the sequence $Y_\mu = [1 - \mu, 1 + \mu] \subset X$ with $0 < \mu < 1$ and $c = c(Y_\mu) = 1$, the d.c. bounding operation yields

$$m(x) = f(c) + f'(c)(x - c) = 3x - 2,$$

and therefore

$$LB(Y_\mu) = 3(1 - \mu) - 2 = 1 - 3\mu$$

because

$$r(Y_\mu) = 1 - \mu.$$

Hence, we obtain

$$\frac{f(r(Y_\mu)) - LB(Y_\mu)}{\delta(Y_\mu)^3} = \frac{1 - 3\mu + 3\mu^2 - \mu^3 - (1 - 3\mu)}{8\mu^3}$$
$$= \frac{3\mu^2 - \mu^3}{8\mu^3} = \frac{3}{8\mu} - \frac{1}{8}.$$

Inasmuch as $3/8\mu$ is unbounded for $\mu \to 0$, we cannot find a fixed constant $C > 0$ such that

$$f(r(Y_\mu)) - LB(Y_\mu) \leq C \cdot \delta(Y_\mu)^3$$

for all $\mu > 0$. Therefore, the d.c. bounding operation does in general not have a rate of convergence of $p \geq 3$.

Remark 3.3. The d.c. bounding operation is a more sophisticated bounding operation with a rate of convergence of $p = 2$. Although the calculation of the lower bounds are again straightforward if a d.c. decomposition is given, it should be mentioned again that the calculation of a d.c. decomposition is in general far from trivial. Furthermore, a d.c. decomposition is not unique; see Section 1.4. Hence, it is a priori unclear which decomposition might yield better results compared to some other ones. Moreover, note that the theoretical rate of convergence for the d.c. bounding operation again strongly depends on the choice of $r(Y)$.

3.4 D.c.m. bounding operation

Recently, Blanquero and Carrizosa (2009) improved the d.c. bounding operation using a d.c.m. decomposition: we rewrite the objective function as a difference of two convex and monotone functions.

Definition 3.1. A function $\varphi : [0,\infty) \to \mathbb{R}$ is called a ***d.c.m. function*** if there exist two convex and monotone functions $\varphi_1, \varphi_2 : [0,\infty) \to \mathbb{R}$ such that

$$\varphi(t) = \varphi_1(t) - \varphi_2(t).$$

Assume that the objective function can be written as

$$f(x) = \varphi(d(x)),$$

where $d : X \to [0,\infty)$ is a nonnegative convex function and further assume that the function $\varphi : [0,\infty) \to \mathbb{R}$ is a d.c.m. function. We find

$$\varphi(t) = \varphi_1(t) - \varphi_2(t),$$

where φ_1 and φ_2 are convex and monotone on $[0,\infty)$.

In order to calculate a lower bound for an arbitrary subbox $Y \subset X \subset \mathbb{R}^n$, Blanquero and Carrizosa (2009) distinguish between two cases for both, for φ_1 and for φ_2.

1. If φ_1 is nonincreasing, define $p = d(c)$ with $c = c(Y)$. Then any subgradient $\xi_1 \in \mathbb{R}$ of φ_1 at p yields

$$\psi_1(t) := \varphi_1(p) + \xi_1(t-p) \leq \varphi_1(t).$$

φ_1 is nonincreasing, thus we have $\xi_1 \leq 0$. Hence,

$$\phi_1(x) := \psi_1(d(x)) = \varphi_1(p) + \xi_1(d(x) - p) \leq \varphi_1(d(x))$$

and ϕ_1 is concave.

2. If φ_1 is nondecreasing, define $\psi_1(x) := \varphi_1(d(x))$. Then ψ_1 is also convex because d is convex and the composition of a nondecreasing convex function with a convex function is convex again. Hence, any subgradient $\xi_1 \in \mathbb{R}^n$ of ψ_1 at $c = c(Y)$ yields

$$\begin{aligned}\phi_1(x) := \psi_1(c) + (\xi_1)^T(x-c) &= \varphi_1(d(c)) + (\xi_1)^T(x-c)\\ \leq \psi_1(x) &= \varphi_1(d(x))\end{aligned}$$

for all $x \in Y$. Note that ϕ_1 is affine linear and therefore concave.

For the functions φ_2, we obtain the following two cases.

1. If φ_2 is nonincreasing, let $\xi_2 \in \mathbb{R}^n$ be any subgradient of d at $c = c(Y)$ and define

$$\psi_2(x) := d(c) + (\xi_2)^T(x-c) \leq d(x).$$

Because φ_2 is nonincreasing, we obtain

$$\phi_2(x) := -\varphi_2(\psi_2(x)) = -\varphi_2\left(d(c) + (\xi_2)^T(x-c)\right) \leq -\varphi_2(d(x))$$

and ϕ_2 is concave because ψ_2 is linear and $-\varphi_2$ is concave.

2. If φ_2 is nondecreasing, define

$$\phi_2(x) := -\varphi_2(d(x)).$$

Because φ_2 is nondecreasing and convex and d is convex, $\varphi_2(d(x))$ is convex again as before. Hence, ϕ_2 is concave.

Summarizing these results, we found

$$m(x) := \phi_1(x) + \phi_2(x) \leq \varphi_1(d(x)) - \varphi_2(d(x)) = \varphi(d(x)) = f(x)$$

for all $x \in Y$ and m is a sum of concave functions and therefore concave again. Denote by $V(Y)$ the set of the 2^n vertices of Y. Then we obtain the ***d.c.m. bounding operation***

$$LB(Y) := \min_{v \in V(Y)} m(v) \quad \text{and} \quad r(Y) \in \arg\min_{v \in V(Y)} m(v). \tag{3.6}$$

The following result shows again a rate of convergence of $p = 2$.

Theorem 3.5. *Consider*

$$f(x) = \varphi_1(d(x)) - \varphi_2(d(x)),$$

where the functions $\varphi_1, \varphi_2 : [0,\infty) \to \mathbb{R}$ are convex, monotone, and twice continuously differentiable on $[0,\infty)$ and $d : X \to [0,\infty)$ is convex and twice continuously differentiable on X.

Then the d.c.m. bounding operation has a rate of convergence of $p = 2$.

Proof. Making use of Lemma 3.1 again, it is sufficient to show

$$f(c) = m(c) \text{ and } \nabla f(c) = \nabla m(c)$$

for all subboxes $Y \subset X$, where $c = c(Y)$. Therefore, we distinguish again between all possible cases.

1. If φ_1 is nonincreasing, we have

$$\phi_1(x) = \varphi_1(p) + (\varphi_1)'(p) \cdot (d(x) - p)$$

with $p = d(c)$. Hence, we find

$$\begin{aligned} \phi_1(c) = \varphi_1(p) &= \varphi_1(d(c)), \\ \nabla\phi_1(c) = (\varphi_1)'(p) \cdot \nabla d(c) &= (\varphi_1)'(d(c)) \cdot \nabla d(c) = \nabla\varphi_1(d(c)), \end{aligned}$$

where we used the generalized chain rule.
2. If φ_1 is nondecreasing, we have

$$\phi_1(x) = \varphi_1(d(c)) + \nabla\varphi_1(d(c))^T \cdot (x - c)$$

and we directly find $\phi_1(c) = \varphi_1(d(c))$ and $\nabla\phi_1(c) = \nabla\varphi_1(d(c))$.

For the functions ϕ_2 we obtain similar results.

1. If φ_2 is nonincreasing, we have

$$\phi_2(x) = -\varphi_2\left(d(c) + \nabla d(c)^T \cdot (x - c)\right)$$

which yields

$$\begin{aligned} \phi_2(c) &= -\varphi_2(d(c)), \\ \nabla\phi_2(c) &= -(\varphi_2)'(d(c)) \cdot \nabla d(c) = -\nabla\varphi_2(d(c)) \end{aligned}$$

using again the generalized chain rule.
2. If φ_2 is nondecreasing, we have $\phi_2(x) = -\varphi_2(d(x))$ and therefore

$$-\varphi_2(d(x)) - \phi_2(x) = 0 \text{ for all } x \in Y.$$

In summary we obtain $m(c) = f(c)$ and $\nabla m(c) = \nabla f(c)$. Hence Lemma 3.1 yields

$$f(r(Y)) - LB(Y) = f(r(Y)) - m(r(Y)) \leq C \cdot \delta(Y)^2$$

for a $C > 0$ that does not depend on Y and the rate of convergence of $p = 2$ is shown.
□

Next, we show that the d.c.m. bounding operation does in general not have a rate of convergence of $p > 2$.

Example 3.4. Consider the convex function $d(x) = x^2$ and the d.c.m. functions $\varphi_1(t) = t$ and $\varphi_2(t) = 0$. Note that φ_1 is convex and nondecreasing on $[0,\infty)$. Thus, the objective function is

$$f(x) = \varphi_1(d(x)) = x^2.$$

Assume the initial box $X = [0,2]$ and the sequence $Y_\mu = [1-\mu, 1+\mu] \subset X$ with $0 < \mu < 1$ and $c = c(Y_\mu) = 1$. Hence, the d.c.m. bounding operation with

$$m(x) = \phi_1(x) = f(c) + f'(c) \cdot (x-1) = 1 + 2 \cdot (x-1) = 2x - 1$$

yields $r(Y_\mu) = 1 - \mu$ and

$$LB(Y_\mu) = m(r(Y_\mu)) = 2(1-\mu) - 1 = 1 - 2\mu.$$

Thus, we obtain

$$\frac{f(r(Y_\mu)) - LB(Y_\mu)}{\delta(Y_\mu)^3} = \frac{(1-\mu)^2 - (1-2\mu)}{8\mu^3} = \frac{1}{8\mu}$$

and we cannot find a fixed constant $C > 0$ such that

$$f(r(Y_\mu)) - LB(Y_\mu) \leq C \cdot \delta(Y_\mu)^3$$

for all $\mu > 0$. Therefore, the d.c.m. bounding operation does in general not have a rate of convergence of $p \geq 3$.

3.4.1 D.c.m. bounding operation for location problems

In Blanquero and Carrizosa (2009), the d.c.m. bounding operation was applied to facility location problems as follows.

Assume that the objective function can be written as

$$f(x) = \sum_{k=1}^{s} \varphi^k(d_k(a_k, x)),$$

where $a_1, \ldots, a_s \in \mathbb{R}^n$ are some given demand points and $d_k(a_k, x)$ for $k = 1, \ldots, s$ are some nonnegative and convex distance functions; for instance, $d_k(a_k, x) = \|x - a_k\|$ for an arbitrary norm $\| \cdot \|$.

Moreover, assume that the functions $\varphi^k : [0,\infty) \to \mathbb{R}$ are d.c.m. functions; that is we find

$$\varphi^k(x) = \varphi_1^k(x) - \varphi_2^k(x)$$

for $k = 1, \ldots, s$, where φ_1^k and φ_2^k are convex and monotone on $[0,\infty)$.

In the same way as before, we can calculate for all $k = 1, \ldots, s$ concave functions ϕ_1^k and ϕ_2^k such that

$$m(x) := \sum_{k=1}^{s} \left(\phi_1^k(x) + \phi_2^k(x) \right) \leq \sum_{k=1}^{s} \left(\varphi_1^k(d_k(a_k,x)) - \varphi_2^k(d_k(a_k,x)) \right) = f(x)$$

for all $x \in Y$. Moreover, note that m is a sum of concave functions and therefore concave. Thus, the d.c.m. bounding operation reads again as

$$LB(Y) := \min_{v \in V(Y)} m(v) \quad \text{and} \quad r(Y) \in \arg\min_{v \in V(Y)} m(v). \tag{3.7}$$

Corollary 3.1. *Consider*

$$f(x) = \sum_{k=1}^{s} \left(\varphi_1^k(d_k(a_k,x)) - \varphi_2^k(d_k(a_k,x)) \right),$$

where the functions $\varphi_1^k, \varphi_2^k : [0,\infty) \to \mathbb{R}$ are convex, monotone, and twice continuously differentiable on $[0,\infty)$ and the distance functions $d_k(a_k,x)$ are convex and twice continuously differentiable on X for $k = 1,\ldots,s$.

Then the d.c.m. bounding operation has a rate of convergence of $p = 2$.

Proof. The proof is analogous to Theorem 3.5 and therefore omitted here. □

Remark 3.4. The d.c.m. bounding operation has again a theoretical rate of convergence of $p = 2$ and the corresponding lower bounds can be derived without difficulties. However, some specific requirements on the objective function are needed and therefore the d.c.m. bounding operation cannot be applied in general. But note that if these requirements are satisfied, then the d.c.m. bounding operation might be much more efficient compared to the d.c. bounding operation as shown in Blanquero and Carrizosa (2009); see also Section 3.10.

3.5 General bounding operation

Our aim in this section is to present a general bounding operation. To this end, we first focus on scalar functions before we consider the general case.

3.5.1 General bounding operation for scalar functions

Consider a compact interval $X \subset \mathbb{R}$ and a scalar function $f : X \to \mathbb{R}$. Moreover, boxes $Y \subset X$ are given by

$$Y = [c - \lambda, c + \lambda] \subset X$$

with $\lambda \geq 0$ and $c = c(Y)$. Assume that f is twice continuously differentiable and that a bounding operation $(L(Y), c(Y))$ of f'' with a rate of convergence of 1 is known. We assume $L(Y) \leq f''(x)$ for all $x \in Y$ with

$$f''(c) - L(Y) \leq D \cdot \delta(Y)$$

for some constant $D > 0$. Note that the lower bound $L(Y)$ depends on Y, but D does not. Defining

$$m(x) := f(c) + f'(c)(x-c) + \frac{1}{2}L(Y)(x-c)^2,$$

we know that

$$f''(x) - m''(x) = f''(x) - L(Y) \geq 0 \text{ for all } x \in Y.$$

Thus, $f - m$ is a convex function on Y; see Theorem 1.1. With $f(c) - m(c) = 0$ and $f'(c) - m'(c) = 0$ the subgradient of $f - m$ at c yields

$$f(x) - m(x) \geq (f(c) - m(c)) + (f'(c) - m'(c)) \cdot (x - c) = 0.$$

Hence, we conclude that $f(x) \geq m(x)$ for all $x \in Y$. Setting $d = -f'(c)/L(Y) + c$ we define

$$LB(Y) := \begin{cases} \min\{m(c-\lambda), m(c+\lambda)\} & \text{if } L(Y) \leq 0, \\ \min\{m(c-\lambda), m(c+\lambda)\} & \text{if } L(Y) > 0 \text{ and } d \notin Y, \\ \min\{m(c-\lambda), m(c+\lambda), m(d)\} & \text{if } L(Y) > 0 \text{ and } d \in Y, \end{cases} \quad (3.8)$$

and obtain $LB(Y) \leq m(x) \leq f(x)$ for all $x \in Y$ because $m'(d) = 0$. Furthermore, we assign

$$r(Y) \in \arg\min\{m(c-\lambda), m(c+\lambda), m(d)\}. \quad (3.9)$$

Because $r(Y) \in \{c-\lambda, c+\lambda, d\}$ we have $LB(Y) = m(r(Y))$. Together we have defined the ***general bounding operation*** of order three for scalar functions; see Figure 3.4.

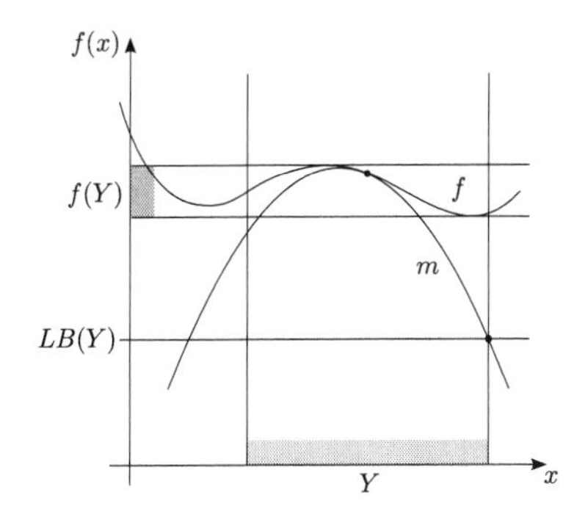

Fig. 3.4 Illustration for the general bounding operation for scalar functions of order three.

Theorem 3.6. *Assume that $f : X \to \mathbb{R}$ is three times continuously differentiable and let a bounding operation $(L(Y), c(Y))$ for f'' with a rate of convergence of 1 be known.*

Then the bounding operation as defined before has a rate of convergence of $p = 3$.

Proof. Using the Taylor expansion of f at $c = c(Y)$, we obtain

$$f(x) \;=\; f(c) + f'(c)(x-c) + \frac{1}{2}f''(c)(x-c)^2 + \frac{1}{6}f^{(3)}(\vartheta(x))(x-c)^3$$

for a $\vartheta(x) \in Y$. Defining

$$\Gamma \;=\; \max\{|f^{(3)}(\vartheta)| \;:\; \vartheta \in X\},$$

we have

$$\begin{aligned} f(x) - m(x) &= \frac{1}{2}\cdot(f''(c) - L(Y))\cdot(x-c)^2 + \frac{1}{6}\cdot f^{(3)}(\vartheta(x))\cdot(x-c)^3 \\ &\le \frac{1}{2}\cdot D\cdot\delta(Y)\cdot\frac{1}{4}\delta(Y)^2 + \frac{1}{6}\Gamma\cdot\frac{1}{8}\delta(Y)^3 \\ &= \left(\frac{1}{8}D + \frac{1}{48}\Gamma\right)\cdot\delta(Y)^3 \end{aligned}$$

for all $x \in Y$. Thus, with $C = (D/8 + \Gamma/48)$ we derived

$$f(r(Y)) - LB(Y) \;=\; f(r(Y)) - m(r(Y)) \;\le\; C\cdot\delta(Y)^3$$

for all boxes Y as stated in the theorem. □

Note that we assumed $r(Y) = c(Y)$ for the bounding operation of f''. However, the theoretical rate of convergence of $p = 3$ is still the same if the bounding operation of f'' uses any other $r(Y) \in Y$ as long as we consider the Taylor expansion of f at $r(Y)$.

In the remainder of this subsection we consider a more general bounding operation for scalar functions. For an odd $p \ge 3$ assume that f is $p-1$ times differentiable and that a bounding operation of $f^{(p-1)}$ using $r(Y) = c = c(Y)$ with a rate of convergence of 1 is known. That is we assume $L(Y) \le f^{(p-1)}(x)$ for all $x \in Y$ with

$$f^{(p-1)}(c) - L(Y) \;\le\; D\cdot\delta(Y).$$

Defining

$$m(x) \;:=\; \sum_{k=0}^{p-2}\frac{1}{k!}\cdot f^{(k)}(c)\cdot(x-c)^k \;+\; \frac{1}{(p-1)!}\cdot L(Y)\cdot(x-c)^{p-1}$$

and using the Taylor expansion of f at c of order $p-2$ with the Lagrange form of the remainder term,

$$f(x) = \sum_{k=0}^{p-2} \frac{1}{k!} \cdot f^{(k)}(c) \cdot (x-c)^k + \frac{1}{(p-1)!} \cdot f^{(p-1)}(\xi(x)) \cdot (x-c)^{p-1}$$

with $\xi(x) \in Y$, we obtain

$$f(x) - m(x) = \frac{1}{(p-1)!} \cdot \underbrace{\left(f^{(p-1)}(\xi(x)) - L(Y)\right)}_{\geq 0} \cdot \underbrace{(x-c)^{p-1}}_{\geq 0} \geq 0 \qquad (3.10)$$

because p is odd. Hence, $f(x) \geq m(x)$ for all $x \in Y$ and we derived the ***general bounding operation*** for scalar functions

$$LB(Y) := \min_{x \in Y} m(x) \quad \text{and} \quad r(Y) \in \arg\min_{x \in Y} m(x). \qquad (3.11)$$

We remark that for any subbox $Y \subset X$ in order to obtain $LB(Y)$ and $r(Y)$ we have to minimize a polynomial m of degree $p-1$.

Theorem 3.7. *Assume that $f : X \to \mathbb{R}$ is p times continuously differentiable for an odd $p \geq 3$ and let a bounding operation $(L(Y), c(Y))$ for $f^{(p-1)}$ with a rate of convergence of 1 be known.*

Then the general bounding operation for scalar functions has a rate of convergence of p.

Proof. The proof is similar to Theorem 3.6 and can also be found in Schöbel and Scholz (2010b). Using the Taylor expansion of f at c of order $p-1$ with the Lagrange form of the remainder term, we obtain

$$f(x) = \sum_{k=0}^{p-1} \frac{1}{k!} \cdot f^{(k)}(c) \cdot (x-c)^k + \frac{1}{p!} \cdot f^{(p)}(\vartheta(x)) \cdot (x-c)^p$$

for a $\vartheta(x) \in Y$. Defining

$$\Gamma = \max\{|f^{(p)}(\vartheta)| : \vartheta \in X\},$$

we have

$$\begin{aligned} f(x) - m(x) &= \frac{1}{(p-1)!} \cdot (f^{(p-1)}(c) - L(Y)) \cdot (x-c)^{p-1} + \frac{1}{p!} \cdot f^{(p)}(\vartheta(x)) \cdot (x-c)^p \\ &\leq \frac{1}{(p-1)!} \cdot D \cdot \delta(Y) \cdot \frac{1}{2^{p-1}} \cdot \delta(Y)^{p-1} + \frac{1}{p!} \Gamma \cdot \frac{1}{2^p} \cdot \delta(Y)^p \\ &= \left(\frac{D}{(p-1)! \cdot 2^{p-1}} + \frac{\Gamma}{p! \cdot 2^p} \right) \cdot \delta(Y)^p \end{aligned}$$

for all $x \in Y$. Thus, with

$$C = \left(\frac{D}{(p-1)! 2^{p-1}} + \frac{\Gamma}{p! 2^p} \right)$$

we derive

$$f(r(Y)) - LB(Y) \;=\; f(r(Y)) - m(r(Y)) \;\leq\; C \cdot \delta(Y)^p$$

as stated in the theorem. □

We remark that a similar bounding operation for $Y = [c-\lambda, c+\lambda]$ for an even $p \geq 2$ can be found if the Taylor expansion of f at $\ell = c - \lambda$ is used. Otherwise we cannot guarantee that $m(x) \leq f(x)$ for all $x \in Y$; see Equation (3.10).

3.5.2 General bounding operation

In this subsection, we derive a general bounding operation with a rate of convergence of p for an arbitrary $p \geq 2$. Therefore, for any box

$$Y \;=\; [y_1^L, y_1^R] \times \cdots \times [y_n^L, y_n^R] \;\subset\; X \;\subset\; \mathbb{R}^n$$

it is necessary to minimize a polynomial $m : \mathbb{R}^n \to \mathbb{R}$ of degree $p-1$. For our calculations, define

$$\ell \;=\; \ell(Y) \;=\; (\ell_1, \ldots, \ell_n) \;=\; (y_1^L, \ldots, y_n^L) \;\in\; Y$$

and suppose that all partial derivatives of f up to order $p-1$ exist and are continuous on X. Furthermore, we assume that some lower bounds $L_{(k_1,\ldots,k_n)}(Y)$ with

$$L_{(k_1,\ldots,k_n)}(Y) \;\leq\; \frac{\partial^{p-1} f}{\partial^{k_1} x_1 \cdots \partial^{k_n} x_n}(x) \;\; \text{for all } x \in Y$$

and for all $k_1, \ldots, k_n \in \mathbb{N}$ with $k_1 + \cdots + k_n = p-1$ are known. Next, consider

$$m(x) = \sum_{k=0}^{p-2} \left(\sum_{\substack{k_1,\ldots,k_n \in \mathbb{N} \\ k_1+\cdots+k_n = k}} \frac{1}{\prod_{i=1}^n k_i!} \cdot \frac{\partial^k f}{\partial^{k_1} x_1 \cdots \partial^{k_n} x_n}(\ell) \cdot \prod_{i=1}^n (x_i - \ell_i)^{k_i} \right)$$
$$+ \sum_{\substack{k_1,\ldots,k_n \in \mathbb{N} \\ k_1+\cdots+k_n = p-1}} \frac{1}{\prod_{i=1}^n k_i!} \cdot L_{(k_1,\ldots,k_n)}(Y) \cdot \prod_{i=1}^n (x_i - \ell_i)^{k_i}.$$

Using the Taylor expansion of f at ℓ of order $p-2$ with the Lagrange form of the remainder term, we obtain for a $\xi(x) \in Y$,

$$f(x) - m(x) \;=$$
$$\sum_{\substack{k_1,\ldots,k_n \in \mathbb{N} \\ k_1+\cdots+k_n = p-1}} \frac{1}{\prod_{i=1}^n k_i!} \cdot \underbrace{\left(\frac{\partial^{p-1} f}{\partial^{k_1} x_1 \cdots \partial^{k_n} x_n}(\xi(x)) - L_{(k_1,\ldots,k_n)}(Y) \right)}_{\geq 0} \cdot \underbrace{\prod_{i=1}^n (x_i - \ell_i)^{k_i}}_{\geq 0}$$

for all $x \in Y$. Thus, we conclude that $f(x) \geq m(x)$ for all $x \in Y$ and we have the ***general bounding operation***

$$LB(Y) := \min_{x \in Y} m(x) \quad \text{and} \quad r(Y) \in \arg\min_{x \in Y} m(x); \tag{3.12}$$

see Figure 3.5.

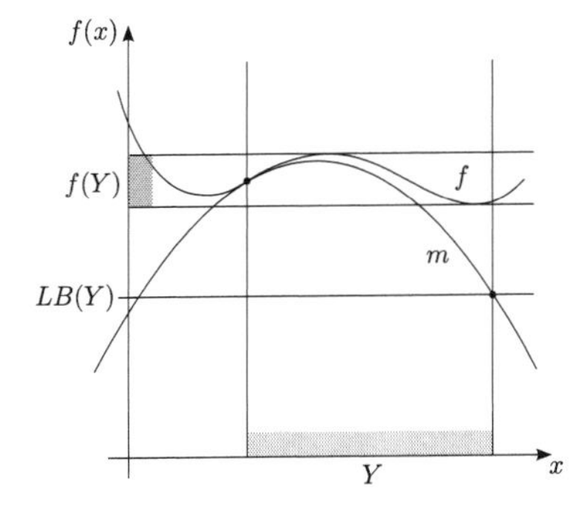

Fig. 3.5 Illustration for the general bounding operation of order three.

Theorem 3.8. *Assume that $f : X \to \mathbb{R}$ is p times continuously differentiable and for all $k_1, \ldots, k_n \in \mathbb{N}$ with $k_1 + \cdots + k_n = p - 1$ let a bounding operation*

$$(L_{(k_1,\ldots,k_n)}(Y),\ \ell(Y))$$

for the function

$$\frac{\partial^{p-1} f}{\partial^{k_1} x_1 \cdots \partial^{k_n} x_n}(x)$$

with a rate of convergence of 1 *be known.*

Then the general bounding operation has a rate of convergence of p.

Proof. In the proof we use the same principles as in our proof of Theorem 3.7. Because

$$(L_{(k_1,\ldots,k_n)}(Y), \ell(Y))$$

are bounding operations with a rate of convergence of 1, there exist constants $D_{(k_1,\ldots,k_n)}$ such that

$$\frac{\partial^{p-1} f}{\partial^{k_1} x_1 \cdots \partial^{k_n} x_n}(\ell) - L_{(k_1,\ldots,k_n)}(Y) \leq D_{(k_1,\ldots,k_n)} \cdot \delta(Y)$$

for all $k_1, \ldots, k_n \in \mathbb{N}$ with $k_1 + \cdots + k_n = p - 1$. The Taylor expansion of f at ℓ of order $p - 1$ with the Lagrange form of the remainder yields

$$f(x) - m(x) =$$

$$\sum_{\substack{k_1,\ldots,k_n \in \mathbb{N} \\ k_1+\cdots+k_n=p-1}} \frac{1}{\prod_{i=1}^n k_i!} \cdot \underbrace{\left(\frac{\partial^{p-1} f}{\partial^{k_1} x_1 \cdots \partial^{k_n} x_n}(\ell) - L_{(k_1,\ldots,k_n)}(Y) \right)}_{\leq D_{(k_1,\ldots,k_n)} \cdot \delta(Y)} \cdot \underbrace{\prod_{i=1}^n (x_i - \ell_i)^{k_i}}_{\leq \delta(Y)^{p-1}}$$

$$+ \sum_{\substack{k_1,\ldots,k_n \in \mathbb{N} \\ k_1+\cdots+k_n=p}} \frac{1}{\prod_{i=1}^n k_i!} \cdot \frac{\partial^p f}{\partial^{k_1} x_1 \cdots \partial^{k_n} x_n}(\vartheta(x)) \cdot \underbrace{\prod_{i=1}^n (x_i - \ell_i)^{k_i}}_{\leq \delta(Y)^p} \leq C \cdot \delta(Y)^p$$

for some $\vartheta(x) \in Y$ and a suitable constant C that does not depend on Y. Hence, we have

$$f(r(Y)) - LB(Y) = f(r(Y)) - m(r(Y)) \leq C \cdot \delta(Y)^p,$$

which proves the theorem. □

We remark again that the general bounding operation requires a solution to the problem

$$\min_{x \in Y} m(x). \tag{3.13}$$

For example, if $p = 3$ then m is a polynomial of degree two. Therefore, the Hessian of m is a symmetric and real-valued matrix $H \in \mathbb{R}^{n \times n}$. Hence, m is concave if and only if H is negative semidefinite and m is convex if and only if H is positive semidefinite.

If H is negative semidefinite, we only have to investigate the 2^n vertices of Y. If H is not negative semidefinite, the minimum of $m(x)$ with $x \in Y$ is attained at a point $x \in Y$ with $\nabla m(x) = 0$ or at a point x on the boundary of Y. Note that we obtain all zeros of $\nabla m(x)$ by solving a system of n linear equations with n variables.

Moreover, for the special case $p = 3$ and $n = 2$ we can derive a finite dominating set of cardinality nine; that is a set of maximal nine points in Y that contains a solution of problem (3.13). One point is the solution of $\nabla m(x) = 0$, four points are the vertices of Y, and we obtain at most four more points, one on each edge of Y.

We remark that similar calculations for the cases $p = 3$ and $n > 2$ also lead to finite dominating sets, but with a higher cardinality than nine, which in turn increases the runtime of the algorithm.

Remark 3.5. The general bounding operation bases on the Taylor expansion of the objective function and bounding operations with a rate of convergence of 1 for some partial derivatives are required. Moreover, if we are using the general bounding operation of order p, we have to minimize for all subboxes Y a polynomial m of degree $p - 1$ to obtain a lower bound. Hence, for $p > 2$ the general bounding operation requires much more computational effort compared to the previous bounding operations. For instance, even for $n = 2$ and $p = 3$ the minimization of m is not a trivial task as discussed before.

3.6 Natural interval bounding operation

In this section as well as in the following ones we assume that we are dealing with functions f such that the corresponding natural interval extensions F exist. Furthermore, we assume that the intervals $F(Y)$ can be computed easily for all subbox $Y \subset X$; see Section 1.5.

Recall that ^{L} indicates the left endpoint of an interval. Then, from the fundamental theorem of interval analysis, see Theorem 1.3 and Figure 3.6, we directly obtain the ***natural interval bounding operation***

$$LB(Y) := F(Y)^L \quad \text{and} \quad r(Y) := c(Y). \tag{3.14}$$

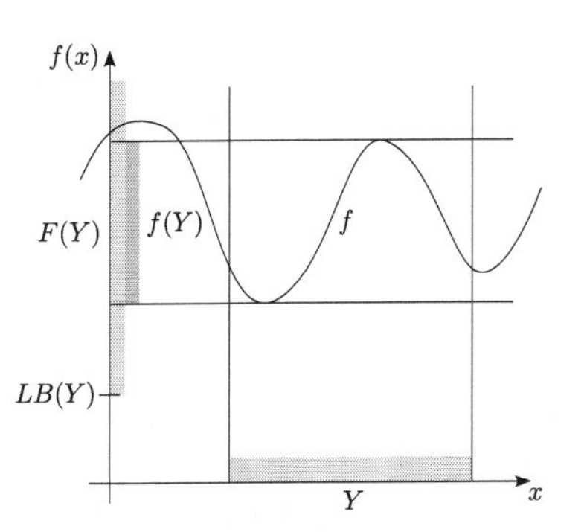

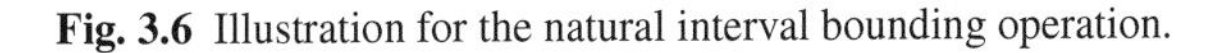

Fig. 3.6 Illustration for the natural interval bounding operation.

Lower bounds derived from the natural interval bounding operation can be used in general for global optimization problems and numerical examples can be found, for instance, in Hansen (1992). Moreover, this bounding operation has been used to solve several location problems; see, for example, Fernández et al. (2007b) and Tóth et al. (2009).

Before we present a general statement concerning the rate of convergence of the natural interval bounding operation, we need the following results.

Lemma 3.2. *Consider a box $X \subset \mathbb{R}^n$, let $f : X \to \mathbb{R}$ be a Lipschitzian function with constant K such that $f(x)$ is a single-use expression, and let $F(X)$ be the natural interval extension of $f(x)$.*

Then we obtain

$$\delta(F(Y)) = F(Y)^R - F(Y)^L \leq K \cdot \delta(Y)$$

for all boxes $Y \subset X$.

Proof. From Theorem 1.4 we know that

$$f(Y) = F(Y) = [F(Y)^L, F(Y)^R]$$

for all subboxes $Y \subset X$ because $f(x)$ is a single-use expression. Thus, there are some $\ell, r \in Y$ with

$$f(\ell) = F(Y)^L \text{ and } f(r) = F(Y)^R.$$

Therefore, we find

$$F(Y)^R - F(Y)^L = f(r) - f(\ell) \leq K \cdot \|r - \ell\|_2 \leq K \cdot \delta(Y)$$

and the statement is shown. □

Theorem 3.9. *Consider a box $X \subset \mathbb{R}^n$ and let $f: X \to \mathbb{R}$ be a Lipschitzian function such that $f(x)$ is a single-use expression.*

Then the natural interval bounding operation has a rate of convergence of $p = 1$.

Proof. Let K be the Lipschitzian constant of f. Then, from Lemma 3.2, we directly obtain

$$f(r(Y)) - LB(Y) = f(c(Y)) - F(Y)^L \leq F(Y)^R - F(Y)^L \leq K \cdot \delta(Y)$$

and the theorem is shown. □

The following example shows that we cannot expect a rate of convergence higher than $p = 1$.

Example 3.5. Consider $f(x) = x^2$ and the box $X = [0, 2]$. Furthermore, define the sequence $Y_\mu = [1 - \mu, 1 + \mu]$ for $0 < \mu < 1$ with $r(Y_\mu) = c(Y_\mu) = 1$. Then the natural interval bounding operation yields

$$LB(Y_\mu) = (Y_\mu^2)^L = (1 - \mu)^2 = 1 - 2\mu + \mu^2.$$

Thus, we find

$$\frac{f(r(Y_\mu)) - LB(Y_\mu)}{\delta(Y_\mu)^2} = \frac{1 - (1 - 2\mu + \mu^2)}{4\mu^2} = \frac{2\mu - \mu^2}{4\mu^2} = \frac{1}{2\mu} - \frac{1}{4}$$

which is unbounded for $\mu \to 0$. Hence, we cannot find a fixed $C > 0$ such that

$$f(r(Y_\mu)) - LB(Y_\mu) \leq C \cdot \delta(Y)^2$$

for all $\mu > 0$.

But even for the following class of functions we can prove a rate of convergence of $p = 1$.

Lemma 3.3. *Consider a box $X \subset \mathbb{R}^n$, let $m_1, \ldots, m_s : X \to \mathbb{R}$ be Lipschitzian functions such that $m_1(x)$ to $m_s(x)$ are single-use expressions, and let $M_1(X)$ to $M_s(X)$*

be the natural interval extensions of $m_1(x)$ to $m_s(x)$, respectively. Furthermore, consider

$$f(x) \;=\; h(m_1(x),\ldots,m_s(x)),$$

where $h:\mathbb{R}^s \to \mathbb{R}$ is a Lipschitzian function such that $h(x)$ is a single-use expression.

Then there exists a fixed constant $C>0$ that does not depend on Y with

$$\delta(F(Y)) \;=\; F(Y)^R - F(Y)^L \;\leq\; C\cdot\delta(Y)$$

for all boxes $Y\subset X$, where $F(X)$ is the natural interval extension of $f(x)$.

Proof. Let $H(X)$ be the natural interval extension of $h(x)$. For any $Y\subset X$, we find some

$$\ell=(\ell_1,\ldots,\ell_s),\; r=(r_1,\ldots,r_s) \;\in\; M_1(Y)\times\cdots\times M_s(Y)$$

such that

$$H(M_1(Y),\ldots,M_s(Y))^L \;=\; h(\ell) \;\text{ and }\; H(M_1(Y),\ldots,M_s(Y))^R \;=\; h(r)$$

inasmuch as $h(x)$ is a single-use expression. Next, let K be the Lipschitzian constant of h. Then, Lemma 3.2 yields

$$\begin{aligned}
\delta(F(Y)) &= \delta(H(M_1(Y),\ldots,M_s(Y)))\\
&= H(M_1(Y),\ldots,M_s(Y))^R - H(M_1(Y),\ldots,M_s(Y))^L\\
&= h(r) - h(\ell) \;\leq\; K\cdot\|r-\ell\|_2\\
&= K\cdot\sqrt{(r_1-\ell_1)^2+\cdots+(r_s-\ell_s)^2}\\
&\leq K\cdot\sqrt{\sum_{k=1}^{s}\left(M_k(Y)^R-M_k(Y)^L\right)^2}\\
&\leq K\cdot\sqrt{\sum_{k=1}^{s}(C_k\cdot\delta(Y))^2} \;\leq\; K\cdot\sqrt{s}\cdot C_{\max}\cdot\delta(Y)
\end{aligned}$$

with $C_{\max}=\max\{C_1,\ldots,C_s\}$. □

Finally, we can prove the following result.

Theorem 3.10. *Consider a box $X\subset\mathbb{R}^n$ and let $m_1,\ldots,m_s: X\to\mathbb{R}$ be Lipschitzian functions such that $m_1(x)$ to $m_s(x)$ are single-use expressions. Furthermore, consider*

$$f(x) \;=\; h(m_1(x),\ldots,m_s(x)),$$

where $h:\mathbb{R}^s\to\mathbb{R}$ is a Lipschitzian function such that $h(x)$ is a single-use expression.

Then the natural interval bounding operation has a rate of convergence of $p=1$.

Proof. Using the previous result, we directly obtain

$$f(r(Y)) - LB(Y) \;\leq\; F(Y)^R - F(Y)^L \;\leq\; C\cdot\delta(Y),$$

which shows the rate of convergence of $p = 1$. □

Remark 3.6. To sum up, we found a general class of functions such that the natural interval bounding operation has a rate of convergence of $p = 1$. We remark that almost all commonly used objective functions for facility location problems can be expressed in the required form. Furthermore, using standard interval analysis tools, the calculation of the lower bounds is again straightforward. Moreover, note that the choice of $r(Y)$ does not affect the rate of convergence; that is for all $r(Y) \in Y$ we obtain a rate of convergence of $p = 1$.

3.7 Centered interval bounding operation

Apart from the natural interval bounding operation we can use the natural interval extension of the Taylor expansion which can also be found in Hansen (1992). We now want to show that the first-order Taylor expansion leads to a bounding operation with a rate of convergence of $p = 2$.

Consider $X \subset \mathbb{R}^n$ and assume that $f : X \to \mathbb{R}$ is continuously differentiable. Then, for all subboxes $Y = Y_1 \times \cdots \times Y_n \subset X$ and $c = (c_1, \ldots, c_n) = c(Y)$ we know that

$$f(x) \;=\; f(c) + \nabla f(\vartheta(x))^T \cdot (x - c) \quad \text{for all } x \in Y$$

and $\vartheta(x) \in Y$. Again using the fundamental theorem of interval analysis, this Taylor expansion leads to

$$f(Y) \;=\; f(Y_1, \ldots, Y_n) \;\subset\; f(c) + \sum_{k=1}^{n} G_k(Y) \cdot (Y_k - c_k),$$

where $G_k(X)$ is the natural interval extension of

$$g_k(x) \;:=\; \frac{\partial f}{\partial x_k}(x) \quad \text{for } k = 1, \ldots, n.$$

Finally, define $z = (z_1, \ldots, z_n) \in Y$ with $z_k \in \{Y_k^L, Y_k^R\}$ such that

$$(G_k(Y) \cdot (Y_k - c_k))^L \;=\; (G_k(Y) \cdot (z_k - c_k))^L$$

for $k = 1, \ldots, n$. Hence, we have constructed the ***centered interval bounding operation***

$$LB(Y) \;:=\; f(c) + \sum_{k=1}^{n} (G_k(Y) \cdot (Y_k - c_k))^L \quad \text{and} \quad r(Y) \;:=\; z. \tag{3.15}$$

Theorem 3.11. *Let $X \subset \mathbb{R}^n$ and consider a continuously differentiable function $f : X \to \mathbb{R}$ such that the natural interval extensions of*

$$g_k(x) = \frac{\partial f}{\partial x_k}(x) \text{ for } k = 1, \ldots, n$$

satisfy the conditions given in Lemma 3.3.

Then the centered interval bounding operation has a rate of convergence of $p = 2$.

Proof. For all $Y \subset X$, the first-order Taylor expansion of f yields

$$f(x) = f(c) + \nabla f(\vartheta(x))^T \cdot (x - c)$$

with $\vartheta(x) \in Y$ for all $x \in Y$. Next, for all $k = 1, \ldots, n$, find $w_k \in G_k(Y)$ such that

$$(G_k(Y) \cdot (Y_k - c_k))^L = w_k \cdot (z_k - c_k),$$

where $z_k \in Y_k$ as defined before. Note that these values exist because $x \cdot y$ is a single-use expression. Finally, define

$$u = (u_1, \ldots, u_n) = \nabla f(\vartheta(z)) \in G_1(Y) \times \cdots \times G_n(Y).$$

Thus, Lemma 3.3 yields

$$\begin{aligned}
f(r(Y)) - LB(Y) &= \nabla f(\vartheta(z))^T \cdot (z - c) - \sum_{k=1}^{n} (G_k(Y) \cdot (Y_k - c_k))^L \\
&= u^T \cdot (z - c) - \sum_{k=1}^{n} w_k \cdot (z_k - c_k) \\
&= \sum_{k=1}^{n} (u_k - w_k) \cdot (z_k - c_k) \\
&\leq \sum_{k=1}^{n} |u_k - w_k| \cdot |z_k - c_k| \\
&\leq \sum_{k=1}^{n} \delta(G_k(Y)) \cdot \frac{1}{2}\delta(Y) \leq \sum_{k=1}^{n} C_k \cdot \delta(Y) \cdot \frac{1}{2}\delta(Y) \\
&= \left(\frac{1}{2} \cdot \sum_{k=1}^{n} C_k\right) \cdot \delta(Y)^2
\end{aligned}$$

and the rate of convergence of $p = 2$ is shown. □

Example 3.6. Consider $f(x) = x^3$, the box $X = [0, 2]$, and the sequence $Y_\mu = [1 - \mu, 1 + \mu]$ for $0 < \mu < 1$ with $c = c(Y_\mu) = 1$. Then we find

$$f(c) + F'(Y_\mu) \cdot (Y_\mu - c) = 1 + 3 \cdot [1 - \mu, 1 + \mu]^2 \cdot [-\mu, \mu].$$

Thus, the centered interval bounding operation yields

$$LB(Y_\mu) = 1 + 3 \cdot \left([(1 - \mu)^2, (1 + \mu)^2] \cdot [-\mu, \mu] \right)^L$$

$$= 1 + 3 \cdot (1+\mu)^2 \cdot (-\mu) \; = \; 1 - 3\mu - 6\mu^2 - 3\mu^3$$

and we obtain $r(Y_\mu) = z = (1-\mu)$. This leads to

$$\begin{aligned}
\frac{f(r(Y_\mu)) - LB(Y_\mu)}{\delta(Y_\mu)^3} &= \frac{(1-\mu)^3 - (1 - 3\mu - 6\mu^2 - 3\mu^3)}{8\mu^3} \\
&= \frac{(1 - 3\mu + 3\mu^2 - \mu^3) - (1 - 3\mu - 6\mu^2 - 3\mu^3)}{8\mu^3} \\
&= \frac{9\mu^2 + 2\mu^3}{8\mu^3} \;=\; \frac{9}{8\mu} + \frac{1}{4},
\end{aligned}$$

which is again unbounded for $\mu \to 0$. Therefore, the centered interval bounding operation has a rate of convergence of not higher than $p = 2$.

Remark 3.7. The centered interval bounding operation is a more sophisticated interval bounding operation with a rate of convergence of $p = 2$. However, note that it requires much more computational effort compared to the natural interval bounding operation because we have to evaluate the natural interval extension of the gradient of the objective function. Moreover, note that the choice of $r(Y)$ is again relevant for the theoretical rate of convergence. Due to this fact, it is also not possible to generalize the centered interval bounding operation to an order of $p > 2$ using a higher-order Taylor expansion.

3.8 Baumann's interval bounding operation

In the previous section, we made use of the first-order Taylor expansion of f at $c = c(Y)$. But note that for both the calculation of the bounding operation and the rate of convergence of $p = 2$, we can use the Taylor expansion of f at any point $y \in Y$. The idea of Baumann (1988) was to use a specific point $b = b(Y) \in Y$ instead of $c = c(Y) \in Y$ as follows.

Consider again a box $X \subset \mathbb{R}^n$ and assume that $f : X \to \mathbb{R}$ is continuously differentiable. Denote by $G_k(X)$ again the natural interval extensions of

$$g_k(x) \;:=\; \frac{\partial f}{\partial x_k}(x) \;\text{ for } k = 1, \ldots, n.$$

Then we define $b = (b_1, \ldots, b_n) \in Y$ by

$$b_k \;=\; \begin{cases} (\delta(G_k(Y))^{-1} \cdot \left(G_k(Y)^R \cdot Y_k^L - G_k(Y)^L \cdot Y_k^R \right) & \text{if } 0 \in G_k(Y) \\ Y_k^L & \text{if } G_k(Y)^L \geq 0 \\ Y_k^R & \text{if } G_k(Y)^R \leq 0 \end{cases}$$

for $k = 1, \ldots, n$; see Baumann (1988). As before, we obtain

$$f(Y) = f(Y_1,\ldots,Y_n) \subset f(b) + \sum_{k=1}^{n} G_k(Y)\cdot(Y_k - b_k).$$

Defining $z = (z_1,\ldots,z_n) \in Y$ with $z_k \in \{Y_k^L, Y_k^R\}$ such that

$$(G_k(Y)\cdot(Y_k - b_k))^L = (G_k(Y)\cdot(z_k - b_k))^L$$

for $k = 1,\ldots,n$, we have ***Baumann's interval bounding operation***

$$LB(Y) := f(b) + \sum_{k=1}^{n}(G_k(Y)\cdot(Y_k - b_k))^L \quad \text{and} \quad r(Y) := z. \tag{3.16}$$

Analogous to the centered interval bounding operation, we obtain quadratic convergence.

Theorem 3.12. *Let $X \subset \mathbb{R}^n$ and consider a continuously differentiable function $f: X \to \mathbb{R}$ such that the natural interval extensions of*

$$g_k(x) = \frac{\partial f}{\partial x_k}(x) \text{ for } k = 1,\ldots,n$$

satisfy the conditions given in Lemma 3.3.

Then Baumann's interval bounding operation has a rate of convergence of $p = 2$.

Proof. The proof is similar to Theorem 3.11 and therefore omitted here. □

Baumann (1988) has shown that the specific choice of b leads to the ***optimal centered form*** in the following sense. For all $y = (y_1,\ldots,y_n) \in Y$ we have

$$f(y) + \sum_{k=1}^{n}(G_k(Y)\cdot(Y_k - y_k))^L \leq f(b) + \sum_{k=1}^{n}(G_k(Y)\cdot(Y_k - b_k))^L.$$

Thus, Baumann's interval bounding operation yields the strongest lower bound among all possible centered forms. Hence, in our numerical results presented in Section 3.10 we expect a smaller constant C compared to the centered interval bounding operation.

We remark that in the literature some related results about quadratic convergence for Baumann's form can be found. For instance, under certain conditions Chuba and Miller (1972) and Krawczyk and Nickel (1982) proved that

$$\delta\left(f(b) + \sum_{k=1}^{n} G_k(Y)\cdot(Y_k - b_k)\right) - \delta(f(Y)) \leq C\cdot\delta(Y)^2.$$

But note that our result is stronger than this one inasmuch as we in general do not have any information about $\delta(f(Y))$ in the left hand-side of this inequality.

Remark 3.8. Baumann's interval bounding operation is similar to the centered interval bounding operation. We are now using the first-order Taylor expansion not at

$c = c(Y)$ but at $b \in Y$. Hence, we obtain the same rate of convergence, but we expect a smaller constant C; see Section 3.10.

3.9 Location bounding operation

In this subsection, we discuss the bounding operation presented in Plastria (1992) for location problems from the point of view of our concept. Let the objective be a function of distances between given demand points and a new single facility location.

Consider s demand points $a_1,\ldots,a_s \in \mathbb{R}^n$ and for each demand point a distance function $d_k(a_k,x)$ for $k = 1,\ldots,s$, for instance, $d_k(a_k,x) = \|x - a_k\|$ for any norm $\|\cdot\|$. Furthermore, we assume that $h : \mathbb{R}^s \to \mathbb{R}$ is a Lipschitzian function such that we are in a position to solve problems of the form

$$\min\{h(z) \ : \ \ell_k \leq z_k \leq u_k \text{ for } k = 1,\ldots,s\},$$

where $\ell_k, u_k \in \mathbb{R}$ are some parameters for $k = 1,\ldots,s$. The objective function for our facility location problem is then given by

$$f(x) \ = \ h(d_1(a_1,x),\ldots,d_s(a_s,x)). \tag{3.17}$$

In order to calculate a lower bound $LB(Y)$ for an arbitrary box $Y \subset \mathbb{R}^n$, suppose that the values

$$d_k^{\min}(Y) = \min\{d_k(a_k,x) \ : \ x \in Y\},$$
$$d_k^{\max}(Y) = \max\{d_k(a_k,x) \ : \ x \in Y\}$$

for $k = 1,\ldots,s$ are easily derived. This is the case if d_k are norms or polyhedral gauges; see Plastria (1992). Obviously, we obtain the ***location bounding operation***

$$LB(Y) := \min\{h(z) \ : \ d_k^{\min} \leq z_k \leq d_k^{\max} \text{ for } k = 1,\ldots,s\}, \tag{3.18}$$
$$r(Y) := c(Y). \tag{3.19}$$

We remark that this bounding operation results in exactly the same lower bound if we use objective functions as given in (3.17) and the natural interval extension from interval analysis, see Section 3.6 and Theorem 3.10 in particular. The main result of Plastria (1992) can now be reformulated within our new concept as follows.

Theorem 3.13. *Consider*

$$f(x) \ = \ h(d_1(a_1,x),\ldots,d_s(a_s,x)),$$

where $h : \mathbb{R}^s \to \mathbb{R}$ is a Lipschitzian function with constant L and $d_1,\ldots,d_s$ are some norms.

Then the location bounding operation has a rate of convergence of $p = 1$.

Proof. This proof can be found in Plastria (1992). Let b_k for $k = 1,\dots,s$ be constants with

$$d_k(x,y) \le b_k \cdot \|x-y\|_2 \text{ for all } x,y \in X,$$

let $b = (b_1,\dots,b_s)$, and define

$$D = [d_1^{\min}, d_1^{\max}] \times \cdots \times [d_s^{\min}, d_s^{\max}] \subset \mathbb{R}^s.$$

One easily sees that

$$|v_k - w_k| \le |d_k^{\max} - d_k^{\min}| \le b_k \cdot \delta(Y)$$

for all $v_k, w_k \in [d_k^{\min}, d_k^{\max}]$ and $k = 1,\dots,s$. These inequalities together yield

$$\|v-w\|_2 \le \|b\|_2 \cdot \delta(Y)$$

for all $v,w \in D \subset \mathbb{R}^s$. Now select $\ell, u \in D$ with $h(\ell) = LB(Y)$ and $u_k = d_k(a_k, c(Y))$ for $k = 1,\dots,s$. We then obtain

$$f(r(Y)) - LB(Y) = h(u) - h(\ell) \le L \cdot \|u-\ell\|_2 \le L \cdot \|b\|_2 \cdot \delta(Y).$$

Using $C = L \cdot \|b\|_2$, we have

$$f(r(Y)) - LB(Y) \le C \cdot \delta(Y)$$

and therefore the rate of convergence of $p = 1$. □

Finally, the next example shows that the location bounding operation cannot be improved to a rate of convergence of $p > 1$.

Example 3.7. Consider the two demand points $a_1 = 0$ and $a_2 = 2$, the initial box $X = [0,2] \subset \mathbb{R}$, and the objective function

$$f(x) = d(a_1,x) + d(a_2,x) = |x| + |2-x|.$$

Then, for the sequence $Y_\mu = [1-\mu, 1+\mu] \subset X$ with $0 < \mu < 1$, the location bounding operation yields the lower bounds

$$LB(Y_\mu) = d_1^{\min}(Y_\mu) + d_2^{\min}(Y_\mu) = (1-\mu) + (1-\mu) = 2(1-\mu).$$

With $r(Y_\mu) = c(Y_\mu) = 1$ we obtain

$$\frac{f(r(Y_\mu)) - LB(Y_\mu)}{\delta(Y_\mu)^2} = \frac{2 - 2(1-\mu)}{4\mu^2} = \frac{1}{2\mu}.$$

This expression is unbounded for $\mu \to 0$, thus we cannot find a fixed constant $C > 0$ such that

$$f(r(Y_\mu)) - LB(Y_\mu) \le C \cdot \delta(Y_\mu)^2.$$

Remark 3.9. As mentioned before, the location bounding operation yields the same lower bound if we use objective functions as given in (3.17) and the natural interval extension. However, both methods differ in the way of constructing the lower bounds. For the location bounding operation we first of all have to calculate the values $d_k^{\min}(Y)$ and $d_k^{\max}(Y)$ for all $k = 1,\ldots,s$, which is not required for the natural interval bounding operation. Finally, note that the rate of convergence of $p = 1$ is again independent of the choice of $r(Y) \in Y$.

3.10 Numerical results

Changing the inequality into an equality in Equation (2.1) in Chapter 2 and applying the natural logarithm yields

$$\log(f(r(Y)) - LB(Y)) = \log(C) + p \cdot \log(\delta(Y)). \tag{3.20}$$

For a given test function f, we can calculate the left-hand side of this expression and $\log(\delta(Y))$ for some selected boxes $Y \subset X$. We then obtain the empirical rate of convergence p and $\log(C)$ by linear regression. This strategy was also used in Tóth and Csendes (2005) and Schöbel and Scholz (2010b) to analyze related concepts.

To present some numerical experiments, we consider the following test function. Assume a given set of m demand points $a_1,\ldots,a_m \in \mathbb{R}^2$ and nonnegative weights $w_1,\ldots,w_m$. Then we define the objective function

$$f(x) = -\sum_{k=1}^{m} w_k \cdot \exp(-\|x - a_k\|_2^2).$$

This objective function is twice continuously differentiable, therefore we could directly apply the natural interval bounding operation, the centered interval bounding operation, and Baumann's interval bounding operation. Note again that the location bounding operation yields exactly the same bounds as the natural interval bounding operation. Moreover, we consider the general bounding operation of order three. To this end, the second partial derivatives are bounded using the natural interval bounding operation. Furthermore, for the d.c. bounding operation Lemma 1.11 easily yields the d.c. decomposition $f = g - h$ with

$$g(x) = \sum_{k=1}^{m} w_k \cdot \|x - a_k\|_2^2,$$

$$h(x) = \sum_{k=1}^{m} w_k \cdot \left(\exp(-\|x - a_k\|_2^2) + \|x - a_k\|_2^2\right).$$

Finally, with $\varphi_k^1(x) = 0$, with $\varphi_k^2(x) = w_k \cdot \exp(-x)$, and with $d_k(a_k, x) = \|x - a_k\|_2^2$ for $k = 1,\ldots,m$ we make use of the d.c.m. bounding operation.

In our numerical experiences, we generated problem instances with $m = 100$ demand points uniformly distributed in $X = [0,10] \times [0,10]$ and weights uniformly distributed in $[0,10]$.

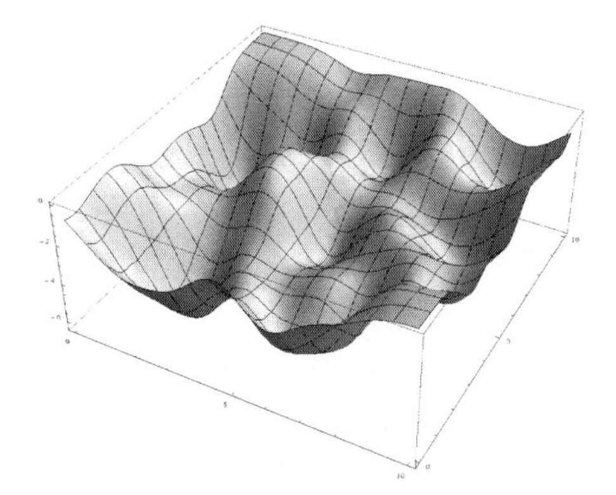

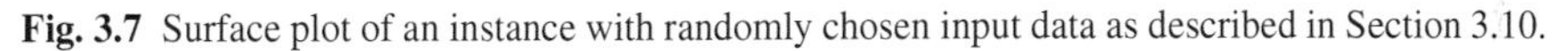

Fig. 3.7 Surface plot of an instance with randomly chosen input data as described in Section 3.10.

The surface plot of one instance with randomly chosen input data is illustrated in Figure 3.7. It is easy to see that the objective function is neither convex nor concave. Hence, geometric branch-and-bound algorithms are suitable solution methods. In the following, three different studies are presented.

3.10.1 Randomly selected boxes

In a first study, we considered one particular problem instance with randomly generated input data as described before. We then selected 200 boxes $Y \subset X$ with different widths such that $f(r(Y)) - LB(Y) \neq 0$ for all bounding operations because $\log(0)$ does not exist; see Equation (3.20).

Our results for these 200 boxes are depicted in Figure 3.8. Linear regressions now yield the values of C and p summarized in Table 3.1.

Bounding Operation	C	p
D.c. bounding operation	128.97	2.00
D.c.m. bounding operation	1.21	2.05
General bounding operation	94.54	3.19
Natural interval bounding operation	2.32	0.97
Centered interval bounding operation	7.91	1.99
Baumann's interval bounding operation	5.34	2.10

Table 3.1 Numerical results for the empirical rate of convergence using randomly selected boxes.

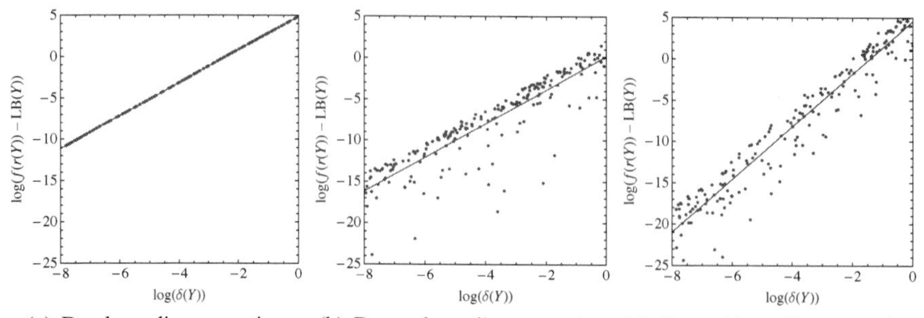

(a) D.c. bounding operation (b) D.c.m. bounding operation (c) General bounding operation

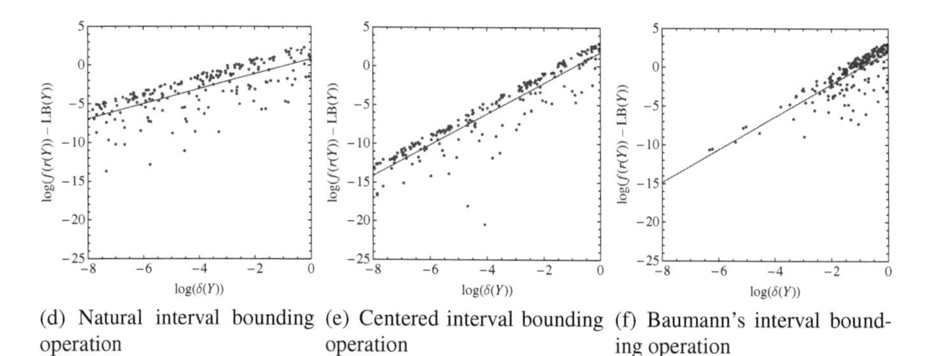

(d) Natural interval bounding operation (e) Centered interval bounding operation (f) Baumann's interval bounding operation

Fig. 3.8 Numerical results for the empirical rate of convergence using randomly selected boxes.

As can be seen, the empirical rates of convergence agree fairly well with our theoretical results. Note that the constant C is smaller for Baumann's interval bounding operation compared to the centered interval bounding operation as expected from Section 3.8. Moreover, note that the variation of our data for the d.c. bounding operation is very small because the Hessian of g does not depend on x; see the proofs of Lemma 3.1 and Theorem 3.4.

3.10.2 Solving one particular problem instance

In our second study, we again used the same problem instance as before but we now considered all boxes that occur while running the geometric branch-and-bound method with $\varepsilon = 10^{-12}$. All selected boxes throughout the algorithm were split into $s = 4$ congruent subboxes. We remark that because the natural interval bounding operation has a rate of convergence of only $p = 1$, the branch-and-bound method is too slow for this bounding operation and it is omitted from our further consideration.

Moreover, the d.c.m. bounding operation $(LB_{\text{dcm}}(Y), r_{\text{dcm}}(Y))$ has the smallest value of C, therefore it might be the best choice for larger boxes. However, the general bounding operation $(LB_{\text{gen}}(Y), r_{\text{gen}}(Y))$ might be the best choice for small boxes because it has the largest value of p. Therefore, we suggest the ***combined bounding operation***

$$\left(\max\{LB_{\text{dcm}}(Y), LB_{\text{gen}}(Y)\},\ \arg\min\{f(r_{\text{dcm}}(Y)),\ f(r_{\text{gen}}(Y))\} \right).$$

We now obtain the results illustrated in Figure 3.9. The empirical rates of convergence are given in Table 3.2.

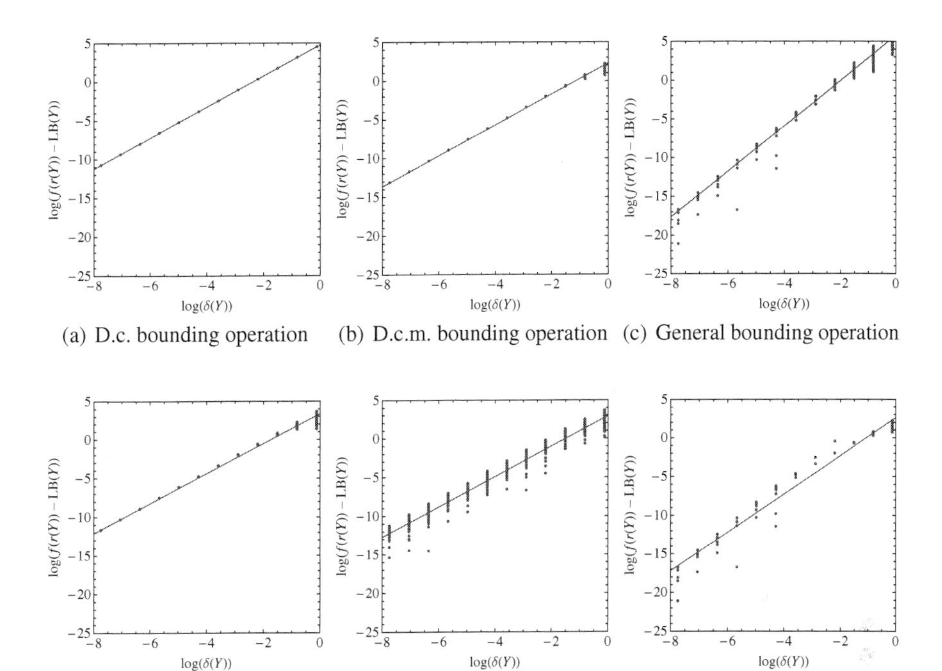

(a) D.c. bounding operation (b) D.c.m. bounding operation (c) General bounding operation

(d) Centered interval bounding operation (e) Baumann's interval bounding operation (f) Combined bounding operation

Fig. 3.9 Numerical results for the empirical rate of convergence solving one particular problem instance.

Bounding Operation	Iterations	C	p
D.c. bounding operation	1,584	118.57	2.00
D.c.m. bounding operation	212	9.73	1.99
General bounding operation	945	327.77	2.93
Centered interval bounding operation	659	28.33	1.93
Baumann's interval bounding operation	346	18.47	1.95
Combined bounding operation	122	13.71	2.47

Table 3.2 Numerical results for the empirical rate of convergence solving one particular problem instance.

As before, the empirical rates of convergence agree fairly well with our theoretical findings, but the constants C for some bounding operations are much higher than for randomly selected boxes. Finally, note that the combined bounding operation needed the fewest number of iterations.

3.10.3 Number of iterations

In a third study, we analyzed the number of iterations needed throughout the branch-and-bound algorithm for ten different randomly generated problem instances with $m = 100$ demand points. Our results can be found in Table 3.3. We remark that one obtains similar results for different values of m.

Bounding Operation	Min	Max	Ave.
D.c. bounding operation	1,316	3,156	2,251.5
D.c.m. bounding operation	186	332	252.0
General bounding operation	719	1,532	1,058.0
Centered interval bounding operation	476	1,018	714.9
Baumann's interval bounding operation	244	612	400.4
Combined bounding operation	88	220	145.3

Table 3.3 Minimum, maximum, and average number of iterations throughout the branch-and-bound algorithm for ten different problem instances.

Although the general bounding operation has a rate of convergence of $p = 3$, the number of iterations is quite high due to the large constant C. Moreover, note that the number of iterations is smaller for Baumann's interval bounding operation compared to the centered interval bounding operation. This can be explained by the smaller constant C as observed before. The smallest number of iterations was again attained for the combined bounding operation.

3.11 Summary

In this chapter, we discussed the most important concepts in geometric branch-and-bound algorithms, namely the calculation of the required lower bounds. Nine different bounding operations were presented and for each of them we proved the theoretical rate of convergence; see Table 3.4. Our theoretical results were justified by some numerical studies given in the previous section.

We remark that similar numerical results can be found using for example the Weber problem on the plane with positive and negative weights; see Tuy et al. (1995) or Drezner and Suzuki (2004). However, compared to the other bounding operations the d.c. bounding operation performs slightly better for the Weber problem than for

Bounding Operation	p	Requirements on the Objective f
Concave bounding operation	∞	concave
Lipschitzian bounding operation	1	Lipschitzian
D.c. bounding operation	2	d.c. decomposition
D.c.m. bounding operation	2	d.c.m. function of a convex function
General bounding operation	p	p times continuously differentiable
Natural interval bounding operation	1	existence of interval extension
Centered interval bounding operation	2	existence of interval extension for ∇f
Baumann's interval bounding operation	2	existence of interval extension for ∇f
Location bounding operation	1	Lipschitzian function of distances

Table 3.4 Summary of the discussed bounding operations. Theoretical rate of convergence p and the requirements on the objective function.

the test function chosen in Section 3.10; see Schöbel and Scholz (2010b) and Scholz (2011b).

Finally we remark that although the combined bounding operation requires the smallest number of iterations throughout the algorithm, it might not be the best choice concerning the runtime of the method because for all boxes throughout the algorithm two lower bounds are calculated. However, it might be fruitful to use a bounding operation with a larger rate of convergence for smaller boxes and a second bounding operation with a smaller rate of convergence for some larger boxes.

In summary, for facility location problems the d.c.m. bounding operation turns out to be one of the most efficient bounding operations; see also Blanquero and Carrizosa (2009).

Chapter 4
Extension for multicriteria problems

Abstract In this chapter, our goal is to generalize the geometric branch-and-bound algorithm to multicriteria optimization problems. To this end, after a short introduction in Section 4.1 we summarize the fundamental definitions and results about multicriteria optimization in Section 4.2. We remark that most of the results in the following sections can also be found in Scholz (2010). However, we introduce the concept of multicriteria bounding operations in Section 4.3 which has not been used therein. Next, in Section 4.4 we prove the convergence of the proposed algorithm as well as properties of the output set. The method is demonstrated on bicriteria facility location problems in Section 4.5 and some numerical results are given in Section 4.6.

4.1 Introduction

In many real-world applications the decision makers have to take several criteria into account which leads to multicriteria problems. For example, power plants, waste dumps, chemical plants, airports, or train stations might be treated as obnoxious facilities due to pollution, offensive smell, or noise pollution. On the other hand, in order to minimize, for example, the transportation costs between the residents or outside suppliers, the new facility should be built close to the inhabitants. Another example is the dosage of drugs or therapies in medicine. Here, one wants to have a fast recovery of the patient whereas the undesirable side effects have to be minimized at the same time. Obviously, these objectives are conflicting with each other.

In order to give a practical example of a multicriteria location problem in detail, suppose the following situation. A new chemical plant should be built and the decision makers consider two criteria for the location of the new facility: the distance from the new plant to the nearest resident is to be maximized due to air pollution but also the total transportation costs from outside supplier to the plant should be minimized.

Location theory is an important application of multicriteria optimization and several models can be found in the literature. Two of the first works in multicriteria location theory are McGinnis and White (1978) and Chalmet et al. (1981) using distance functions induced by rectilinear norms. Hansen and Thisse (1981) proposed a solution strategy for the bicriteria Weber-Rawls problem and Hamacher and Nickel (1996) considered planar location problems where the objective functions are either Weber or center problems. Today, a wide range of multicriteria location problems with different objectives have been reported; see, for example, Nickel et al. (1997), Brimberg and Juel (1998a,b), Melachrinoudis and Xanthopulos (2003), Blanquero and Carrizosa (2002), Skriver and Anderson (2003), Ohsawa (2000), Ohsawa and Tamura (2003), Ohsawa et al. (2006), or Ohsawa et al. (2008). See Carrizosa and Plastria (1999) for a survey about semiobnoxious facility location problems.

In this chapter, our goal is to generalize the geometric branch-and-bound algorithm to multicriteria optimization problems.

4.2 Multicriteria optimization

In this section, we briefly summarize basic definitions for multicriteria optimization problems with p objectives following Ehrgott (2005).

Definition 4.1. A ***multicriteria*** optimization problem can be formulated as

$$\operatorname{vec}\min_{x\in X} f(x) \ := \ (f_1(x),\dots,f_p(x)), \tag{4.1}$$

where $X \subset \mathbb{R}^n$ is the feasible set and the objectives are $f_i(x) : \mathbb{R}^n \to \mathbb{R}$ for $i = 1,\dots,p$. Define $Y = f(X) \subset \mathbb{R}^p$ as the set of attainable outcomes.

A multicriteria optimization problem with two objective functions is called a ***bicriteria*** optimization problem.

Definition 4.2. Let $x = (x_1,\dots,x_p)$ and $y = (y_1,\dots,y_p)$ be two vectors in $\mathbb{R}^p$.

1. We write $x \leqq y$ if and only if $x_i \leq y_i$ for $i = 1,\dots,p$.
2. We write $x \leq y$ if and only if $x_i \leq y_i$ for $i = 1,\dots,p$ and $x \neq y$.
3. We write $x < y$ if and only if $x_i < y_i$ for $i = 1,\dots,p$.

Denote by $\mathbb{R}^p_{\geqq} = \{x \in \mathbb{R}^p \ : \ x \geqq 0\}$ the set of all vectors in $\mathbb{R}^p$ with nonnegative components. $\mathbb{R}^p_{\geq}$ and $\mathbb{R}^p_{>}$ are defined analogously.

In general there exists no point $x^* \in X$ with $f(x^*) \leqq f(x)$ for all $x \in X$, therefore the following definition presents the concept of Pareto optimal solutions in multicriteria optimization problems.

Definition 4.3. A solution $\hat{x} \in X$ is called ***Pareto optimal*** or ***efficient*** if there is no $x \in X$ satisfying

$$f(x) \leq f(\hat{x}).$$

The set of all Pareto optimal solutions is denoted by $X_E \subset X$ and the ***nondominated set*** is $Y_N = f(X_E)$.

$\hat{x} \in X$ is called ***weakly Pareto optimal*** or ***weakly efficient*** if there is no $x \in X$ satisfying

$$f(x) < f(\hat{x}).$$

The set of all weakly Pareto optimal solutions is denoted by $X_{wE} \subset X$ and the ***weakly nondominated set*** is $Y_{wN} = f(X_{wE})$.

All Pareto optimal solutions are weakly Pareto optimal; that is $X_E \subset X_{wE}$.

In practice, it might be difficult to calculate the complete set X_E and thus we are interested in an ε-approximation of X_E. Two of the first works in this area are White (1986) and Loridan (1984). Therein, a general concept of ε-Pareto optimal solutions is introduced that finds applications in several works; see, for example, Engau and Wiecek (2007a).

Definition 4.4. For a given $\varepsilon \in \mathbb{R}^p_{\geqq}$, a solution $\hat{x} \in X$ is called ***ε-Pareto optimal*** or ***ε-efficient*** if there is no $x \in X$ satisfying

$$f(x) + \varepsilon \leq f(\hat{x}).$$

The set of all ε-Pareto optimal solutions is denoted by X_E^ε and the ε-nondominated set is $Y_N^\varepsilon = f(X_E^\varepsilon)$; see Figure 4.1.

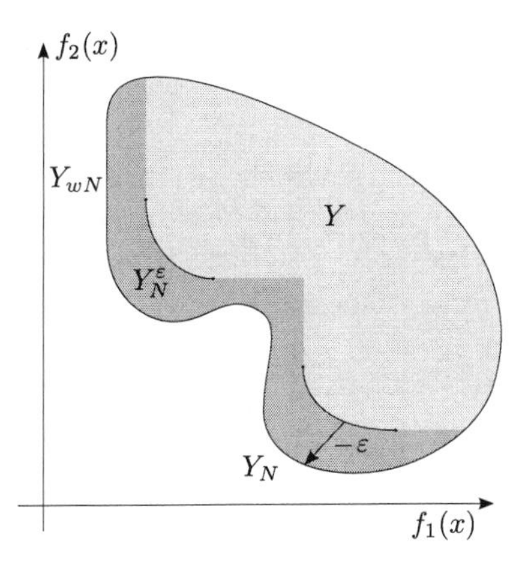

Fig. 4.1 For the definition of ε-Pareto optimal solutions. The dark grey area including the two thin black curves indicates the ε-nondominated set Y_N^ε.

Obviously, we have $X_E \subset X_E^\varepsilon$ for all $\varepsilon \geqq 0$ and $X_E = X_E^0$ for $0 \in \mathbb{R}^p$. The following alternative definition illustrates the concept of ε-Pareto optimal solutions; see Figure 4.2.

Corollary 4.1. *A solution $\hat{x} \in X$ is ε-Pareto optimal if and only if*

$$((f(\hat{x}) - \varepsilon) - \mathbb{R}^p_{\geq}) \cap Y = \emptyset.$$

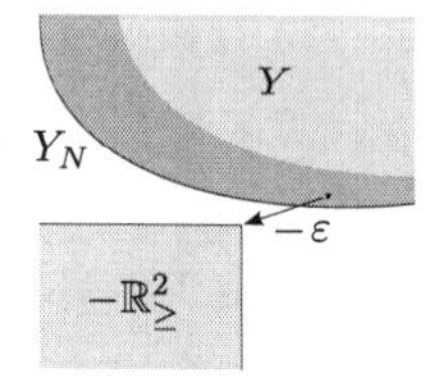

Fig. 4.2 For the alternative definition of ε-Pareto optimal solutions.

In Engau and Wiecek (2007b), seven scalarization methods for the calculation of ε-Pareto optimal solutions are given. The idea of the suggested procedure therein is to find ε-optimal solutions for single objective problems that are ε-Pareto optimal for the corresponding multicriteria optimization problem and ε depends on ε as well as on the chosen scalarization method. We remark that several single objective problems have to be solved. Therefore, the method does not result in a superset of the Pareto optimal set X_E in general. In contrast to this, the idea of the approach given in the next section is to find a subset X_A^ε of X that consists of ε-Pareto optimal solutions and contains all Pareto optimal solutions; that is find a set X_A^ε of X such that

$$X_E \subset X_A^\varepsilon \subset X_E^\varepsilon.$$

Recently, Fernández and Tóth (2007, 2009) suggested outer approximation methods for the Pareto optimal set of bicriteria problems using interval analysis. In their first work, they used regions of δ-optimality of a finite number of single objective constrained problems. Their second work was a more general framework similar to our algorithm with a discussion of the quality of the output set. They used a concept of superdominated points that is related to the concept of ε-efficiency. For a bicriteria optimization problem a solution $\hat{x}$ is ε-Pareto optimal if and only if $\hat{x}$ is not ε-superdominated by any other point.

4.3 The algorithm

Before we can present the general branch-and-bound algorithm, we need to extend the concept of bounding operations to the multicriteria case as follows.

Definition 4.5. Let $X \subset \mathbb{R}^n$ be a box and consider $f : X \to \mathbb{R}^p$. A ***multicriteria bounding operation*** is a procedure to calculate for any subbox $Y \subset X$ lower and upper bounds $LB_i(Y), UB_i(Y) \in \mathbb{R}$ such that

$$LB_i(Y) \leq f_i(x) \leq UB_i(Y)$$

for all $i = 1, \ldots, p$ and all $x \in Y$. Moreover, with $c = c(Y)$ define

$$\begin{aligned} LB(Y) &:= (LB_1(Y), \ldots, LB_p(Y)), \\ UB(Y) &:= (UB_1(Y), \ldots, UB_p(Y)), \\ OV(Y) &:= (f_1(c), \ldots, f_p(c)). \end{aligned}$$

Formally, we obtain the multicriteria bounding operation

$$(LB(Y),\ UB(Y),\ OV(Y))$$

for all subboxes $Y \subset X$.

For the following branch-and-bound method, consider a multicriteria optimization problem

$$\text{vec}\min_{x \in X} f(x) = (f_1(x), \ldots, f_p(x)),$$

where the feasible area is given by a box $X \subset \mathbb{R}^n$. Notice that the algorithm is also suitable for more general shapes of the feasible area X, for example, in the case that the feasible area can be approximated by a union of boxes as before. As input for the algorithm choose an arbitrary $\varepsilon \in \mathbb{R}^p_>$ and assume that a multicriteria bounding operation is known.

1. Let $\mathscr{X}$ be a list of boxes and initialize $\mathscr{X} := \{X\}$.
2. Apply the bounding operations to X; that is calculate

$$LB(X), \ UB(X), \text{ and } OV(X).$$

3. If $LB(X) \leq UB(X) - \varepsilon$, set $T(X) = 1$. Else set $T(X) = 0$.
4. If $T(Y) = 0$ for all $Y \in \mathscr{X}$, the algorithm stops and the output is

$$X_A^{\varepsilon} := \bigcup_{Y \in \mathscr{X}} Y \subset X \subset \mathbb{R}^n.$$

   Else set

$$\delta_{\max} := \max\{\delta(Y) : Y \in \mathscr{X}\}.$$

5. Select a box $Y \in \mathscr{X}$ with $\delta(Y) = \delta_{\max}$ and split it into s congruent smaller subboxes Y_1 to Y_s; see Subsection 2.3.2.
6. Set $\mathscr{X} = (\mathscr{X} \setminus Y) \cup \{Y_1, \ldots, Y_s\}$; that is delete Y from $\mathscr{X}$ and add Y_1 to Y_s.
7. Apply the bounding operations to Y_1 to Y_s; that is for $k = 1, \ldots, s$ calculate

$$LB(Y_k), \ UB(Y_k), \text{ and } OV(Y_k).$$

8. For $k = 1, \ldots, s$, if $LB(Z) \leq UB(Y_k) - \varepsilon$ for some $Z \in \mathscr{X}$, set $T(Y_k) = 1$. Else set $T(Y_k) = 0$.
9. Moreover, if

$$UB(Y) - LB(Y) \leqq \frac{\varepsilon}{2}$$

   for all $Y \in \mathscr{X}$, set $T(Y) = 0$ for all $Y \in \mathscr{X}$.
10. Discard all boxes $Z \in \mathscr{X}$ with $LB(Z) \geq OV(Y_k)$ for some $k \in \{1, \ldots, s\}$. Furthermore, for $k = 1, \ldots, s$ discard Y_k from $\mathscr{X}$ if $LB(Y_k) \geq OV(Z)$ for some $Z \in \mathscr{X}$.
11. Return to Step 4.

In the following section, we show that all $x \in Y$ are ε-Pareto optimal if we set $T(Y) = 0$ throughout the algorithm; see Theorem 4.3. Furthermore, note that Step 9 is only needed to ensure the termination of the method; see Theorem 4.1.

4.4 Convergence theory

In this section, we discuss the convergence of the algorithm as well as properties of the output set. To this end, let us introduce the following definition.

Definition 4.6. Let $X \subset \mathbb{R}^n$ be a box and $f : X \to \mathbb{R}^p$. Furthermore, consider the multicriteria optimization problem

$$\operatorname{vec}\min_{x \in X} f(x) = (f_1(x), \ldots, f_p(x)).$$

We say a multicriteria bounding operation is ***convergent*** if there exists a fixed constant $C > 0$ such that

$$\|UB(Y) - LB(Y)\|_1 \leq C \cdot \delta(Y)$$

for all boxes $Y \subset X$.

A link between the rate of convergence for scalar bounding operations and convergent multicriteria bounding operations yields the following result. Recall that $c(Y)$ is the center of a box Y.

Lemma 4.1. *Let $X \subset \mathbb{R}^n$ be a box and $f : X \to \mathbb{R}^p$. Moreover, for $i = 1, \ldots, p$ assume bounding operations $(LB_i(Y), c(Y))$ for f_i and $(NB_i(Y), c(Y))$ for $-f_i$ with a rate of convergence of 1; that is assume*

$$f_i(c(Y)) - LB_i(Y) \leq C_i \cdot \delta(Y) \text{ and } -f_i(c(Y)) - NB_i(Y) \leq D_i \cdot \delta(Y)$$

for all boxes $Y \subset X$. Finally, define $UB_i(Y) = -NB_i(Y)$. Then

$$(LB(Y),\ UB(Y),\ OV(Y))$$

is a convergent multicriteria bounding operation for

$$\operatorname{vec} \min_{x \in X} f(x) = (f_1(x), \ldots, f_p(x)).$$

Proof. First of all note that we have $LB_i(Y) \leq f_i(x)$ and $NB_i(Y) \leq -f_i(x)$ for all $x \in Y$ by definition of bounding operations and therefore

$$LB_i(Y) \leq f_i(x) \leq -NB_i(Y) = UB_i(Y)$$

for $i = 1, \ldots, p$ and all $x \in Y$. Hence, we indeed have a multicriteria bounding operation. Next, we find

$$\begin{aligned} UB_i(Y) - LB_i(Y) &\leq -NB_i(Y) + C_i \cdot \delta(Y) - f_i(c(Y)) \\ &\leq D_i \cdot \delta(Y) + f_i(c(Y)) + C_i \cdot \delta(Y) - f_i(c(Y)) \\ &= (D_i + C_i) \cdot \delta(Y). \end{aligned}$$

Together, we obtain

$$\begin{aligned} \|UB(Y) - LB(Y)\|_1 &= \sum_{i=1}^{p} \Big(UB_i(Y) - LB_i(Y) \Big) \\ &\leq \sum_{i=1}^{p} \Big((D_i + C_i) \cdot \delta(Y) \Big) = \left(\sum_{i=1}^{p} (D_i + C_i) \right) \cdot \delta(Y) \end{aligned}$$

which shows that the multicriteria bounding operation is convergent. □

To sum up, if we find for all p objective functions lower and upper bounds such that the corresponding bounding operations using $r(Y) = c(Y)$ have a rate of convergence of 1, then the corresponding multicriteria bounding operation is convergent.

The next theorem shows that the proposed algorithm terminates after a finite number of iterations if we assume a convergent multicriteria bounding operation.

Theorem 4.1. *Consider the multicriteria geometric branch-and-bound algorithm with a convergent multicriteria bounding operation. Furthermore, assume that each selected box throughout the algorithm is split into $s = 2^n$ congruent smaller boxes.*

Then the algorithm terminates after a finite number of iterations for every $\varepsilon \in \mathbb{R}^p_>$.

Proof. The multicriteria bounding operation is convergent, therefore we have

$$\|UB(Y) - LB(Y)\|_1 \le C \cdot \delta(Y).$$

for a fixed constant C that does not depend on Y. Thus, we find

$$UB_i(Y) - LB_i(Y) \le C \cdot \delta(Y) \le \frac{\varepsilon_i}{2}$$

for $i = 1, \ldots, p$ and for all boxes $Y \in \mathscr{X}$ after a finite number of iterations because in Step 4 of the algorithm a box with largest diameter is selected for a split into $s = 2^n$ smaller subboxes. Hence, we obtain

$$UB(Y) - LB(Y) \leqq \frac{\varepsilon}{2}$$

for all $Y \in \mathscr{X}$ after a finite number of iterations. Finally, Step 9 ensures the termination of the proposed algorithm. □

As stated above, we now want to show that the algorithm results in the set $X_A^\varepsilon \subset X$ such that

$$X_E \subset X_A^\varepsilon \subset X_E^\varepsilon.$$

For the understanding of the following two theorems, Figure 4.3 shows an example with six boxes in the objective space for a bicriteria problem.

Theorem 4.2. *Let X_A^ε be the output set obtained by the proposed algorithm. Then we have $X_E \subset X_A^\varepsilon$.*

Proof. We have to show that the algorithm does not delete any Pareto optimal solution; that is we have to analyze Step 10.

A box $Y \in \mathscr{X}$ will be discarded if $LB(Y) \ge OV(Z)$ for some $Z \in \mathscr{X}$. In this case, with $c = c(Z) \in Z$ we find

$$f(c) = OV(Z) \le LB(Y) \leqq f(x)$$

for all $x \in Y$. Hence no $x \in Y$ can be Pareto optimal because it is dominated by $c \in Z \subset X$; see Figure 4.3(a). □

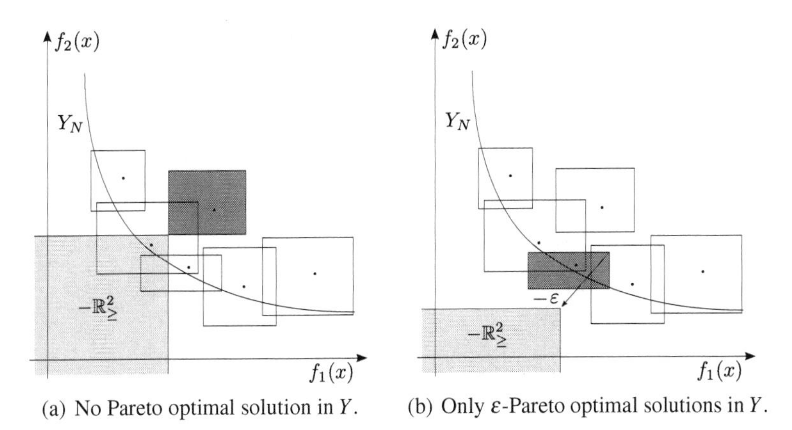

(a) No Pareto optimal solution in Y. (b) Only ε-Pareto optimal solutions in Y.

Fig. 4.3 Some boxes Y with upper bounds $UB(Y)$, lower bounds $LB(Y)$, and objective values $OV(Y)$ in the objective space for a bicriteria problem; that is $p = 2$. (a) An example of a box that contains no Pareto optimal solution. (b) An example of a box that contains only ε-Pareto optimal solutions.

Theorem 4.3. *Let X_A^ε be the output set obtained by the proposed algorithm. Then we have $X_A^\varepsilon \subset X_E^\varepsilon$.*

Proof. The algorithm stops if $T(Y) = 0$ for all $Y \in \mathscr{X}$. Thus, we have to show that all $y \in Y$ are ε-Pareto optimal solutions when we set $T(Y) = 0$ in Step 3, Step 8, or Step 9 of the algorithm.

To this end, let $Y \in \{Y_1, \dots, Y_s\}$ be one of the new boxes in a fixed iteration and assume $\mathscr{X}$ to be the list of boxes after Step 7 in the same iteration. Suppose there is a $y \in Y$ that is not ε-Pareto optimal; that is there exists an $x \in X$ with

$$f(x) + \varepsilon \leq f(y).$$

We distinguish two cases to show that then there is always a $Z \in \mathscr{X}$ with

$$LB(Z) \leq UB(Y) - \varepsilon.$$

1. If $x \notin W$ for all $W \in \mathscr{X}$ then there is a $Z \in \mathscr{X}$ with $LB(Z) \leq f(x)$; compare Step 10. Now we have

$$LB(Z) \leq f(x) \leq f(y) - \varepsilon \leqq UB(Y) - \varepsilon.$$

2. If $x \in Z$ for a $Z \in \mathscr{X}$, we directly obtain

$$LB(Z) \leqq f(x) \leq f(y) - \varepsilon \leqq UB(Y) - \varepsilon.$$

Because we found in both cases a $Z \in \mathscr{X}$ with

$$LB(Z) \leq UB(Y) - \varepsilon,$$

the algorithm ensures that we set $T(Y) = 1$ if Y contains any $y \in Y$ that is not ε-Pareto optimal. In other words, if we set $T(Y) = 0$ all $y \in Y$ are ε-Pareto optimal solutions; see Figure 4.3(b).

Next, assume $\mathscr{X}$ to be the list of boxes after Step 9 in the first iteration with

$$UB(Y) - LB(Y) \leqq \frac{\varepsilon}{2} \tag{4.2}$$

for all $Y \in \mathscr{X}$; see the proof of Theorem 4.1. Suppose there is a $y \in Y$ for a $Y \in \mathscr{X}$ that is not ε-Pareto optimal. Along the lines of the first part of this proof we then find a $Z \in \mathscr{X}$ with

$$LB(Z) \leq UB(Y) - \varepsilon.$$

Using Equation (4.2) for Z and Y, we obtain

$$OV(Z) - \frac{\varepsilon}{2} \leqq UB(Z) - \frac{\varepsilon}{2} \leqq LB(Z) \leq UB(Y) - \varepsilon \leqq LB(Y) - \frac{\varepsilon}{2}$$

and therefore $OV(Z) \leq LB(Y)$. Hence, if we set $T(Y) = 0$ for all $Y \in \mathscr{X}$ in Step 9, the discarding test in Step 10 ensures that at the end of the same iteration all $y \in Y$ for all $Y \in \mathscr{X}$ are ε-Pareto optimal solutions. □

4.5 Example problems

In order to present some numerical results, we consider bicriteria facility location problems. To this end, consider m demand points $a_k \in \mathbb{R}^2$ with weights $w_k, v_k \geq 0$ for $k = 1, \ldots, m$. We treat two planar 1-facility location problems: one new facility $x \in \mathbb{R}^2$ should be built; see also Section 1.3.

4.5.1 Semiobnoxious location problem

In our first example problem we want to minimize the sum of service costs from each demand point to the new facility location. However, we also want to minimize the sum of the reciprocal squared distance from the demand points to the new facility. This problem can be modeled as a biobjective optimization problem

$$\text{vec}\min_{x \in \mathbb{R}^2} (f_1(x), f_2(x)),$$

where the objective functions using the Euclidean norm are given as

$$f_1(x) = \sum_{k=1}^{m} w_k \cdot \|x - a_k\|_2 \text{ and } f_2(x) = \sum_{k=1}^{m} \frac{v_k}{\max\{\|x - a_k\|_2^2, \varepsilon\}},$$

where $\varepsilon > 0$ is a small number. This ***semiobnoxious location problem*** was studied in Brimberg and Juel (1998a) providing a solution approach using the weighted sum method which turned out to be unsuitable for this problem because the second objective function is not convex. Skriver and Anderson (2003) solved one instance of this problem with an algorithm similar to ours but without any criteria for ε-Pareto optimality.

The first objective function is convex, thus lower bounds are easily obtained using the d.c. bounding operation. Furthermore, upper bounds can be constructed using the concave bounding operation because $-f_1$ is concave; see Theorem 3.2. The second objective function can be bounded from below and above using the location bounding operation which yields the same results as the natural interval bounding operation.

4.5.2 Semidesirable location problem

Our second example problem is similar to the first one. One wants to minimize the sum of service costs from each demand point to the new facility location but at the same time one wants to maximize the distance between the new facility and the nearest resident. A model for this problem is given by

$$\text{vec}\min_{x\in\mathbb{R}^2} (f_1(x), f_2(x)),$$

where the objective functions are given as

$$f_1(x) = \sum_{k=1}^{m} w_k \cdot \|x - a_k\|_2 \text{ and } f_2(x) = \max_{1\leq k\leq m} -v_k \cdot \|x - a_k\|_2 .$$

This ***semidesirable location problem*** was analyzed by Brimberg and Juel (1998b) and Melachrinoudis and Xanthopulos (2003) with a solution technique founded on Voronoi diagrams. Ohsawa and Tamura (2003) suggested a solution algorithm using the rectilinear norm for the first objective function and elliptic distances for the second one. Also Blanquero and Carrizosa (2002) considered a generalization of this problem. Their solution method results in a finite set of ε-efficient solutions using a scalarization method.

In our numerical studies, the first objective function was bounded as before. Bounds for f_2 were again obtained using the location bounding operation.

4.6 Numerical results

Before we present general numerical results, we analyze one particular instance of the semiobnoxious location problem.

Example 4.1. Consider the semiobnoxious location problem with $m = 10$ demand points and the input data given in Table 4.1.

k	1	2	3	4	5	6	7	8	9	10
a_k	(2,3)	(7,1)	(8,9)	(2,5)	(6,6)	(4,9)	(9,3)	(4,3)	(3,1)	(1,8)
w_k	30	96	85	92	84	28	4	31	83	74
v_k	57	36	98	34	25	59	27	71	11	60

Table 4.1 Input data for Example 4.1.

To apply the multicriteria geometric branch-and-bound method, we used the feasible region and initial square $X = [0,10] \times [0,10]$ and the vector ε was calculated as in our computational experiences; see below.

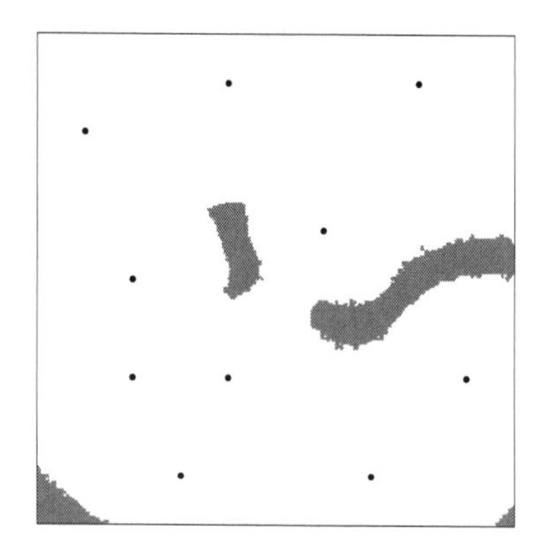

Fig. 4.4 Demand points (black dots) and output set of the algorithm (grey area) for Example 4.1.

For this problem instance, the algorithm terminates after 2,038 iterations. The output set X_A^ε is illustrated in Figure 4.4. Recall that X_A^ε contains all Pareto optimal solutions and all points in X_A^ε are ε-Pareto optimal.

For some further computational experiences, we generated $10 \leq m \leq 10{,}000$ demand points $a_1, \ldots, a_m$ uniformly distributed in $X = [0,1] \times [0,1]$ and weights $w_k, v_k \in [1,5]$ for $k = 1, \ldots, m$. Ten problems were run for different values of m.

We chose the values for $\varepsilon = (\varepsilon_1, \varepsilon_2)$ as follows. For both objective functions we calculated ε-optimal solutions x_1^* and x_2^* for the corresponding single objective problems using the standard geometric branch-and-bound method as given in Chapter 2. We then defined

$$\varepsilon_1 = 0.04 \cdot (f_1(x_2^*) - f_1(x_1^*)) \text{ and } \varepsilon_2 = 0.04 \cdot (f_2(x_1^*) - f_2(x_2^*)).$$

This choice is useful inasmuch as the ranges

$$f_1(X) = \{f_1(x) : x \in X\} \text{ and } f_2(X) = \{f_2(x) : x \in X\}$$

and therefore also $f_1(x_2^*) - f_1(x_1^*)$ and $f_2(x_1^*) - f_2(x_2^*)$ strongly depend on the objective functions $f_1(x)$ and $f_2(x)$. Thus, a fixed ε for all problem instances is not suitable.

Our results for the semiobnoxious location model can be found in Table 4.2 and for the semidesirable location model in Table 4.3. Therein, the last column (Size) indicates the average size of the remaining region in percent in comparison to the initial box.

	Runtime (Sec.)			Iterations			Size (%)
m	Min	Max	Ave.	Min	Max	Ave.	Ave.
10	0.00	0.55	0.13	181	3,657	1,312.8	12.60
20	0.01	0.93	0.21	449	3,413	1,555.2	9.23
50	0.06	1.27	0.48	939	5,294	2,632.4	5.54
100	0.17	5.29	1.41	1,629	9,681	4,332.3	3.11
200	1.45	10.55	3.83	5,512	13,492	8,302.5	1.25
500	0.62	18.86	4.99	1,972	15,256	6,514.3	1.33
1,000	2.80	18.06	8.77	4,290	14,430	9,456.0	0.65
2,000	6.94	26.21	14.66	6,057	16,213	10,422.9	0.64
5,000	20.66	81.43	43.37	7,803	22,330	14,322.9	0.35
10,000	69.57	149.11	117.25	13,302	26,273	20,987.5	0.14

Table 4.2 Numerical results for the semiobnoxious location model.

	Runtime (Sec.)			Iterations			Size (%)
m	Min	Max	Ave.	Min	Max	Ave.	Ave.
10	0.02	0.23	0.09	554	1,401	881.7	2.14
20	0.02	0.26	0.12	508	1,796	1,178.7	1.56
50	0.04	0.36	0.14	711	2,329	1,233.5	0.71
100	0.07	0.70	0.18	855	2,050	1,178.6	0.36
200	0.15	0.47	0.26	995	1,915	1,366.8	0.21
500	0.36	0.91	0.59	1,114	2,538	1,639.5	0.08
1,000	0.68	1.91	1.14	972	2,763	1,721.2	0.03
2,000	2.17	3.58	2.73	1,741	2,858	2,180.7	0.02
5,000	7.33	13.36	9.51	2,385	4,250	3,060.0	0.02
10,000	22.37	35.86	27.71	3,731	5,823	4,580.7	0.02

Table 4.3 Numerical results for the semidesirable location model.

As can be seen, both problems could be solved in a reasonable amount of time. In particular, all problem instances were solved in less than two minutes of computer time. Notice that the average number of iterations increases and the size of the remaining region decreases with the number of demand points for both problems, which might be a consequence of the particular choice of ε.

Chapter 5
Multicriteria discarding tests

Abstract Under the assumption that the objective functions for a multicriteria optimization problem are differentiable, this chapter presents some general discarding tests that can be used throughout the algorithm presented in the previous chapter. The idea of these discarding tests is to obtain a sharp outer approximation of the set of Pareto optimal solutions. To this end, we recall the well-known Fritz John necessary conditions for Pareto optimality in Section 5.1 before the multicriteria discarding tests are presented in Section 5.2. The theoretical results are again illustrated on two bicriteria location problems introduced in Section 5.3. Some particular instances for these problems are solved in Section 5.4 twice, one time without multicriteria discarding tests and one time using these tests. We show that the second run yields a very sharp outer approximation of the set of all Pareto optimal solutions compared to the first run.

5.1 Necessary conditions for Pareto optimality

In this section, we recall the well-known Fritz John necessary conditions for Pareto optimality which can be found, for example, in Da Cunha and Polak (1967) or Miettinen (1999). Therefore, assume that the feasible region X is given by

$$X = \{x \in \mathbb{R}^n : r(x) = (r_1(x), \ldots, r_q(x)) \leqq 0\}$$

for some $r_1, \ldots, r_q : \mathbb{R}^n \to \mathbb{R}$.

Theorem 5.1 (Fritz John conditions for Pareto optimality). *Let $f_1, \ldots, f_p : \mathbb{R}^n \to \mathbb{R}$, let $r_1, \ldots, r_q : \mathbb{R}^n \to \mathbb{R}$, and define*

$$X = \{x \in \mathbb{R}^n : r(x) = (r_1(x), \ldots, r_q(x)) \leqq 0\}.$$

Moreover, assume that $f_1, \ldots, f_p$ and $r_1, \ldots, r_q$ are continuously differentiable at $\hat{x} \in X$. A necessary condition for $\hat{x}$ to be Pareto optimal for

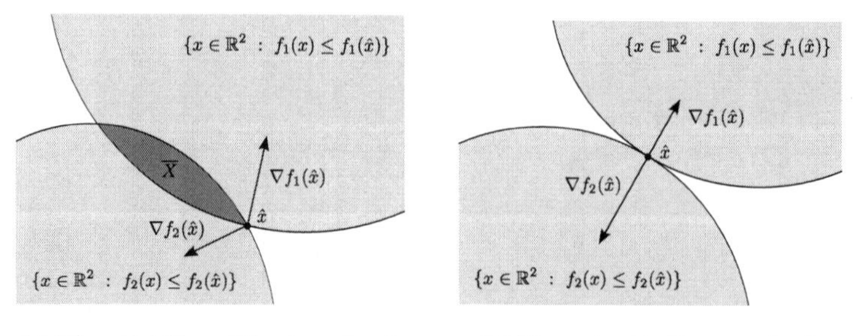

(a) The point $\hat{x}$ is not Pareto optimal. (b) The point $\hat{x}$ is Pareto optimal.

Fig. 5.1 Illustration of the Fritz John conditions for Pareto optimality for an unconstrained bicriteria problem with two variables; that is $n = p = 2$. (a) The point $\hat{x}$ is not Pareto optimal because all $x \in \overline{X}$ dominate $\hat{x}$: $f(x) \le f(\hat{x})$ for all $x \in \overline{X}$. (b) The point $\hat{x}$ satisfies the Fritz John conditions for Pareto optimality: $\nabla f_1(\hat{x}) = -\lambda \cdot \nabla f_2(\hat{x})$ for a $\lambda > 0$.

$$\operatorname{vec}\min_{x\in X} f(x) \;=\; (f_1(x),\ldots,f_p(x))$$

is that there exist a $\lambda \in \mathbb{R}^p_{\geqq}$ *and a* $\mu \in \mathbb{R}^q_{\geqq}$ *with* $(\lambda,\mu) \neq (0,0)$ *such that*

$$\sum_{i=1}^{p} \lambda_i \cdot \nabla f_i(\hat{x}) + \sum_{j=1}^{q} \mu_j \cdot \nabla r_j(\hat{x}) \;=\; 0 \text{ and } \mu_j \cdot r_j(\hat{x}) \;=\; 0$$

for $j = 1,\ldots,q$.

Proof. See, for instance, Da Cunha and Polak (1967). □

We remark that some similar necessary conditions for the case that the objective functions and the constraints are strictly pseudoconvex or nondifferentiable can be found in Majumdar (1997) or Kanniappan (1983), respectively. Some sufficient conditions for differentiable and convex problems are given, for example, in Singh (1987) and Miettinen (1999).

Next, consider a subset $Y \subset X$ with $r(x) < 0$ for all $x \in Y$. Then we conclude that a necessary condition for $\hat{x} \in Y$ to be Pareto optimal is that there exists a $\lambda \in \mathbb{R}^p_{\geq}$ such that

$$\sum_{i=1}^{p} \lambda_i \cdot \nabla f_i(\hat{x}) \;=\; 0. \tag{5.1}$$

Note that $0 \notin \mathbb{R}^p_{\geq}$. Sets Y with $r(x) < 0$ for all $x \in Y$ are of our interest in the following, thus for a number of corollaries we assume $X = \mathbb{R}^n$ only for simplicity; see also Figure 5.1.

Corollary 5.1. *Let* $f_1,\ldots,f_p : \mathbb{R}^n \to \mathbb{R}$ *and* $\hat{x} \in X = \mathbb{R}^n$. *Assume that* $f_1,\ldots,f_p$ *are continuously differentiable at* $\hat{x}$. *If there is an* $s \in \{1,\ldots,n\}$ *with*

$$\frac{\partial f_1}{\partial x_s}(\hat{x}),\ldots,\frac{\partial f_p}{\partial x_s}(\hat{x}) < 0 \text{ or } \frac{\partial f_1}{\partial x_s}(\hat{x}),\ldots,\frac{\partial f_p}{\partial x_s}(\hat{x}) > 0, \tag{5.2}$$

then $\hat{x}$ is not Pareto optimal for the multicriteria optimization problem

$$\text{vec}\min_{x\in\mathbb{R}^n} f(x) = (f_1(x),\ldots,f_p(x)).$$

Proof. If condition (5.2) is satisfied, then there is no $\lambda \in \mathbb{R}^p_{\geq}$ with

$$\sum_{i=1}^{p} \lambda_i \cdot \frac{\partial f_i}{\partial x_s}(\hat{x}) = 0.$$

Therefore, the Fritz John conditions are not satisfied and hence $\hat{x}$ cannot be Pareto optimal. □

Corollary 5.2. *Let $f_1,\ldots,f_p : \mathbb{R}^n \to \mathbb{R}$ and $\hat{x} \in X = \mathbb{R}^n$. Assume that $f_1,\ldots,f_p$ are continuously differentiable at $\hat{x}$. If the vectors*

$$\{\nabla f_1(\hat{x}), \ldots, \nabla f_p(\hat{x})\}$$

are linearly independent then $\hat{x}$ is not Pareto optimal for the multicriteria optimization problem

$$\text{vec}\min_{x\in\mathbb{R}^n} f(x) = (f_1(x),\ldots,f_p(x)).$$

Proof. If $\{\nabla f_1(\hat{x}),\ldots,\nabla f_p(\hat{x})\}$ are linearly independent, Equation (5.1) is not satisfied for any $\lambda \in \mathbb{R}^p \setminus \{0\}$ and hence $\hat{x}$ cannot be Pareto optimal. □

A direct consequence of Corollary 5.2 is the following result.

Corollary 5.3. *Let $f_1,\ldots,f_p : \mathbb{R}^n \to \mathbb{R}$ and $\hat{x} \in X = \mathbb{R}^n$. Assume that $f_1,\ldots,f_p$ are continuously differentiable at $\hat{x}$ and $n \geq p$. If there is a $(p \times p)$-submatrix B of*

$$Df(\hat{x}) = \begin{pmatrix} \frac{\partial f_1}{\partial x_1}(\hat{x}) & \cdots & \frac{\partial f_1}{\partial x_n}(\hat{x}) \\ \vdots & & \vdots \\ \frac{\partial f_p}{\partial x_1}(\hat{x}) & \cdots & \frac{\partial f_p}{\partial x_n}(\hat{x}) \end{pmatrix} \in \mathbb{R}^{p\times n}$$

with $\det(B) \neq 0$ *then $\hat{x}$ is not Pareto optimal for the multicriteria optimization problem*

$$\text{vec}\min_{x\in\mathbb{R}^n} f(x) = (f_1(x),\ldots,f_p(x)).$$

Proof. If there is a $(p \times p)$-submatrix B of $Df(\hat{x})$ with $\det(B) \neq 0$ then the vectors

$$\{\nabla f_1(\hat{x}), \ldots, \nabla f_p(\hat{x})\}$$

are linearly independent. Hence, the statement is a direct consequence of Corollary 5.2. □

We remark that for the case $p = n$ we only have to consider the determinant of

$$Df(\hat{x}) = \begin{pmatrix} \frac{\partial f_1}{\partial x_1}(\hat{x}) & \dots & \frac{\partial f_1}{\partial x_n}(\hat{x}) \\ \vdots & & \vdots \\ \frac{\partial f_n}{\partial x_1}(\hat{x}) & \dots & \frac{\partial f_n}{\partial x_n}(\hat{x}) \end{pmatrix} \in \mathbb{R}^{n \times n}.$$

In location theory, we can find several location models designed as bicriteria optimization problems on the plane; see Section 4.1. Therefore, our next corollary considers the special case of bicriteria optimization problems on the plane: $n = p = 2$.

Corollary 5.4. *Let $f_1, f_2 : \mathbb{R}^2 \to \mathbb{R}$ and $\hat{x} \in X = \mathbb{R}^2$. Assume that f_1 and f_2 are continuously differentiable at $\hat{x}$. If*

$$\frac{\partial f_1}{\partial x_1}(\hat{x}) \cdot \frac{\partial f_2}{\partial x_2}(\hat{x}) - \frac{\partial f_1}{\partial x_2}(\hat{x}) \cdot \frac{\partial f_2}{\partial x_1}(\hat{x}) \neq 0 \text{ or} \tag{5.3}$$

$$\frac{\partial f_1}{\partial x_1}(\hat{x}) \cdot \frac{\partial f_2}{\partial x_1}(\hat{x}) + \frac{\partial f_1}{\partial x_2}(\hat{x}) \cdot \frac{\partial f_2}{\partial x_2}(\hat{x}) > 0, \tag{5.4}$$

then $\hat{x}$ is not Pareto optimal for the bicriteria optimization problem

$$\text{vec} \min_{x \in \mathbb{R}^2} f(x) = (f_1(x), f_2(x)).$$

Proof. If (5.3) is satisfied, the vectors $\{\nabla f_1(\hat{x}), \nabla f_2(\hat{x})\}$ are linearly independent and Corollary 5.2 yields the result.

On the other hand, if (5.3) is not satisfied but (5.4) is satisfied, $\{\nabla f_1(\hat{x}), \nabla f_2(\hat{x})\}$ are linearly dependent and we have $\nabla f_i(\hat{x}) \neq 0$ for $i = 1, 2$ with

$$\nabla f_1(\hat{x}) = \alpha \cdot \nabla f_2(\hat{x})$$

for a unique $\alpha > 0$. Thus, Equation (5.1) reduces to

$$(\alpha \lambda_1 + \lambda_2) \cdot \nabla f_2(\hat{x}) = 0$$

which has no solution $\lambda \in \mathbb{R}^2_{\geq}$ because $\nabla f_2(\hat{x}) \neq 0$ and $\alpha \lambda_1 + \lambda_2 > 0$ for all $\lambda \in \mathbb{R}^2_{\geq}$.
□

5.2 Multicriteria discarding tests

Using the necessary conditions for Pareto optimality presented in the previous section, we now want to derive some discarding tests for the multicriteria branch-and-bound methods suggested in Chapter 4.

Recall that the goal of multicriteria discarding tests is to obtain a sharp outer approximation of the set of all Pareto optimal solutions X_E. In other words, if we can show that a box Y throughout the algorithm does not contain any Pareto optimal solution, the box can be discarded in Step 10 of the algorithm.

To this end, we assume that we are in a position to calculate lower bounds $G(Y)^L$ and upper bounds $G(Y)^R$ on any box Y for some function $g : \mathbb{R}^n \to \mathbb{R}$; that is to calculate real values $G(Y)^L, G(Y)^R \in \mathbb{R}$ such that

$$G(Y)^L \le g(x) \le G(Y)^R \text{ for all } x \in Y.$$

Of course, this can be done using interval analysis, see Section 1.5, because the natural interval extension $G(X) = G(X_1, \dots, X_n)$ of $g(x) = g(x_1, \dots, x_n)$ leads to

$$g(Y) = \{g(x) : x \in Y\} \subset G(Y) = [G(Y)^L, G(Y)^R];$$

see Theorem 1.3. Under these assumptions, we obtain the following discarding tests.

Lemma 5.1. *Let $f_1, \dots, f_p : \mathbb{R}^n \to \mathbb{R}$, consider a box $Y \subset \mathbb{R}^n$, and assume that $f_1, \dots, f_p$ are continuously differentiable at x for all $x \in Y$. Furthermore, assume*

$$\frac{\partial f_i}{\partial x_k}(Y) \subset [G_k^i(Y)^L, G_k^i(Y)^R]$$

for all $i = 1, \dots, p$ and $k = 1, \dots, n$. If there is an $s \in \{1, \dots, n\}$ with

$$G_s^1(Y)^L, \dots, G_s^p(Y)^L > 0 \text{ or } G_s^1(Y)^R, \dots, G_s^p(Y)^R < 0$$

then Y does not contain any Pareto optimal solution for the multicriteria optimization problem

$$\operatorname{vec} \min_{x \in \mathbb{R}^n} f(x) = (f_1(x), \dots, f_p(x)).$$

Proof. Follows directly from Corollary 5.1. □

Lemma 5.2. *Let $f_1, f_2 : \mathbb{R}^2 \to \mathbb{R}$, consider a box $Y \subset \mathbb{R}^2$, assume that f_1 and f_2 are continuously differentiable at x for all $x \in Y$, and define*

$$g(x) := \frac{\partial f_1}{\partial x_1}(x) \cdot \frac{\partial f_2}{\partial x_2}(x) - \frac{\partial f_1}{\partial x_2}(x) \cdot \frac{\partial f_2}{\partial x_1}(x),$$

$$h(x) := \frac{\partial f_1}{\partial x_1}(x) \cdot \frac{\partial f_2}{\partial x_1}(x) + \frac{\partial f_1}{\partial x_2}(x) \cdot \frac{\partial f_2}{\partial x_2}(x).$$

Furthermore, assume that $g(Y) \subset [G(Y)^L, G(Y)^R]$ and $h(Y) \subset [H(Y)^L, H(Y)^R]$. If

$$G(Y)^L > 0, \ G(Y)^R < 0, \text{ or } H(Y)^L > 0,$$

then Y does not contain any Pareto optimal solution for the bicriteria optimization problem

$$\operatorname{vec} \min_{x \in \mathbb{R}^2} f(x) = (f_1(x), f_2(x)).$$

Proof. Follows directly from Corollary 5.4. □

We remark that for bicriteria problems some discarding tests were also presented in Fernández and Tóth (2009). In particular, the first discarding test therein is exactly the same as given in Lemma 5.1 for bicriteria problems and their second one is related to Lemma 5.2. However, note that our discarding test in Lemma 5.2 is more general because it is also suitable for multicriteria problems with $p > 2$.

Lemma 5.3. *Let $f_1, f_2, f_3 : \mathbb{R}^3 \to \mathbb{R}$, consider a box $Y \subset \mathbb{R}^3$, assume that f_1, f_2, and f_3 are continuously differentiable at x for all $x \in Y$, and define*

$$g(x) := \det \begin{pmatrix} \frac{\partial f_1}{\partial x_1}(x) & \dots & \frac{\partial f_1}{\partial x_3}(x) \\ \vdots & & \vdots \\ \frac{\partial f_3}{\partial x_1}(x) & \dots & \frac{\partial f_3}{\partial x_3}(x) \end{pmatrix}.$$

Furthermore, assume that $g(Y) \subset [G(Y)^L, G(Y)^R]$. If

$$G(Y)^L > 0 \text{ or } G(Y)^R < 0,$$

then Y does not contain any Pareto optimal solution for the multicriteria optimization problem

$$\operatorname{vec}\min_{x \in \mathbb{R}^3} f(x) = (f_1(x), f_2(x), f_3(x)).$$

Proof. The statement follows directly from Corollary 5.3. □

Finally, we present a multicriteria discarding test for general multicriteria optimization problems that relies on Corollary 5.3.

Lemma 5.4. *Let $f_1, \dots, f_p : \mathbb{R}^n \to \mathbb{R}$, $n \geq p$, consider a box $Y \subset \mathbb{R}^n$, assume that $f_1, \dots, f_p$ are continuously differentiable at x for all $x \in Y$, and define*

$$g_i(x) := \det(B_i) \text{ for } i = 1, \dots, \binom{n}{p},$$

where B_i are the $(p \times p)$-submatrices of $Df(x)$. Furthermore, assume that

$$g_i(Y) \subset [G_i(Y)^L, G_i(Y)^R]$$

for $i = 1, \dots, \binom{n}{p}$. If there is an $s \in \{1, \dots, \binom{n}{p}\}$ with

$$G_s(Y)^L > 0 \text{ or } G_s(Y)^R < 0$$

then Y does not contain any Pareto optimal solution for the multicriteria optimization problem

$$\operatorname{vec}\min_{x \in \mathbb{R}^n} f(x) = (f_1(x), \dots, f_p(x)).$$

To sum up, all multicriteria discarding tests suggested in this section are collected in Table 5.1.

Discarding Test	Requirements on n and p
Lemma 5.1	$n \geq 1$ and $p \geq 1$
Lemma 5.2	$n = 2$ and $p = 2$
Lemma 5.3	$n = 3$ and $p = 3$
Lemma 5.4	$n \geq 1$ and $p \leq n$

Table 5.1 Summary of the suggested multicriteria discarding tests.

5.3 Example problems

We again want to solve two bicriteria facility location problems on the plane to illustrate the efficiency of the presented discarding tests. To this end, consider again m demand points $a_k \in \mathbb{R}^2$ with weights $w_k, v_k \geq 0$ for $k = 1, \ldots, m$ for the following problems.

5.3.1 Semiobnoxious location problem

Our first example problem is again the semiobnoxious location problem as introduced in Section 4.5.1. We want to solve

$$\operatorname{vec}\min_{x \in \mathbb{R}^2} (f_1(x), f_2(x)),$$

where the objective functions are given as

$$f_1(x) = \sum_{k=1}^{m} w_k \cdot \|x - a_k\|_2 \text{ and } f_2(x) = \sum_{k=1}^{m} \frac{v_k}{\max\{\|x - a_k\|_2^2, \varepsilon\}}$$

with a small number $\varepsilon > 0$. In the following section, both objective functions are bounded from below and above using the location bounding operation.

5.3.2 Bicriteria Weber problem

The second example problem is the bicriteria Weber problem: the Weber problem as presented in Section 1.3 with two objective functions using different weights. Hence, we obtain the bicriteria problem

$$\operatorname{vec}\min_{x \in \mathbb{R}^2} (f_1(x), f_2(x)),$$

where

$$f_1(x) = \sum_{k=1}^{m} w_k \cdot \|x - a_k\|_2 \text{ and } f_2(x) = \sum_{k=1}^{m} v_k \cdot \|x - a_k\|_2 .$$

The bicriteria Weber problem was analyzed in Hamacher and Nickel (1996) and Nickel (1997). Therein, apart from general theoretical results, solution methods are discussed for the rectilinear norm. In our following numerical studies we again used the location bounding operations for both objective functions throughout the branch-and-bound algorithm.

5.4 Numerical examples

For the example problems given in the previous section, we directly obtained discarding tests from Lemmas 5.1 and 5.2. In the following problem instances, we applied the multicriteria branch-and-bound algorithm as given in Chaper 4 twice. In a first run, we did not apply any multicriteria discarding test and we used the discarding tests derived from Lemmas 5.1 and 5.2 in a second run. In both cases, the algorithm were stopped after 10,000 iterations even if the termination criteria as discussed in Chapter 4 was satisfied before. All bounds for the discarding tests were calculated using the natural interval extension; that is the natural interval bounding operation. Moreover, note that we could only apply the discarding tests if both objectives were continuously differentiable for all $x \in Y$; that is for all boxes $Y \subset X$ with $a_k \notin Y$ for $k = 1, \ldots, m$.

Example 5.1. In our first problem instance we consider the semiobnoxious location problem with the same input data as given in Example 4.1; see Table 4.1. The output set of the algorithm for both runs is depicted in Figure 5.2.

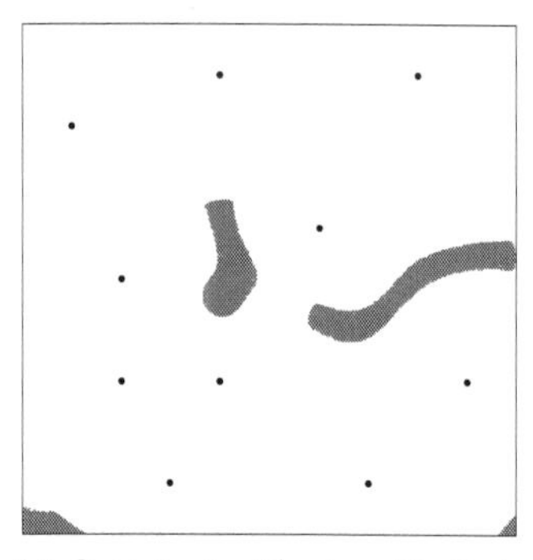

(a) Output set without multicriteria discarding tests.

(b) Output set with multicriteria discarding tests.

Fig. 5.2 Results for Example 5.1. (a) Output set after 10,000 iterations without multicriteria discarding tests. (b) Output set after 10,000 iterations using the multicriteria discarding tests derived from Lemmas 5.1 and 5.2.

Indeed, the multicriteria discarding tests yield a much sharper outer approximation of the set of all Pareto optimal solutions X_E than without these tests. In particular, 20,285 out of the overall number of 40,000 boxes throughout the algorithm could be deleted by a multicriteria discarding test.

Example 5.2. We again consider the semiobnoxious location problem but now with the input data adopted from Brimberg and Juel (1998a); see Table 5.2. The output set for both runs can be seen in Figure 5.3. Here, 19,960 out of the 40,000 boxes throughout the algorithm could be deleted by a multicriteria discarding test.

k	1	2	3	4	5	6	7
a_k	(5,20)	(18,8)	(22,16)	(14,17)	(7,2)	(5,15)	(12,4)
w_k	5	7	2	3	6	1	5
v_k	1	1	1	1	1	1	1

Table 5.2 Input data for Example 5.2.

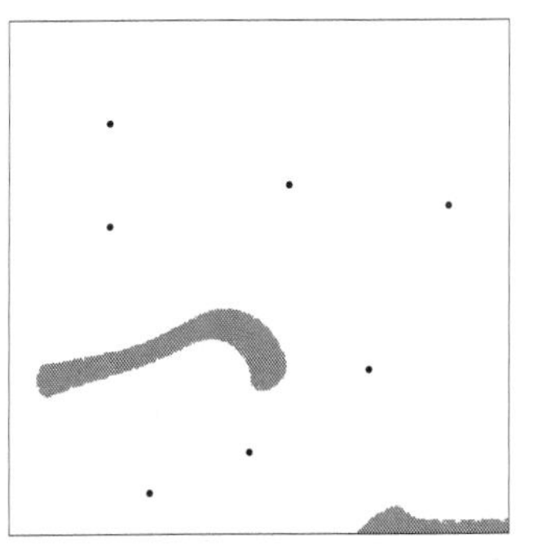

(a) Output set without multicriteria discarding tests.

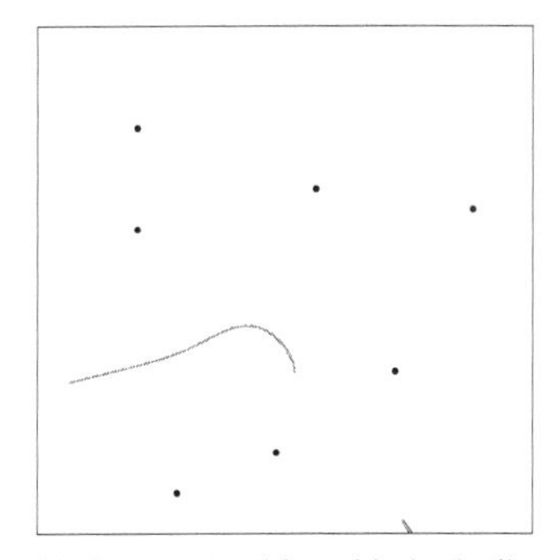

(b) Output set with multicriteria discarding tests.

Fig. 5.3 Results for Example 5.2. (a) Output set after 10,000 iterations without multicriteria discarding tests. (b) Output set after 10,000 iterations using the multicriteria discarding tests derived from Lemmas 5.1 and 5.2.

Example 5.3. Finally, we consider an instance of the bicriteria Weber problem with 12 demand points as given in Table 5.3. The output sets can be found in Figure 5.4. Again, the multicriteria discarding tests yield a much sharper approximation of X_E than without any multicriteria tests inasmuch as 20,041 out of the 40,000 boxes could be deleted.

We remark that for a fixed number of iterations the runtime of the algorithm of course increases using some further discarding tests. However, in our examples the runtimes increased only by a factor of roughly four.

Furthermore, note that some more numerical results can be found in Scholz (2011a). Therein, multicriteria optimization problems with more than $p = 2$ objective functions are also considered and it is shown that the further discarding tests

k	1	2	3	4	5	6	7	8	9	10	11	12
a_k	(1,5)	(2,9)	(2,1)	(3,2)	(4,1)	(5,8)	(6,6)	(5,4)	(7,1)	(8,6)	(9,6)	(9,3)
w_k	2	5	15	11	24	5	11	30	1	1	73	88
v_k	9	12	82	56	60	49	20	49	39	8	19	5

Table 5.3 Input data for Example 5.3.

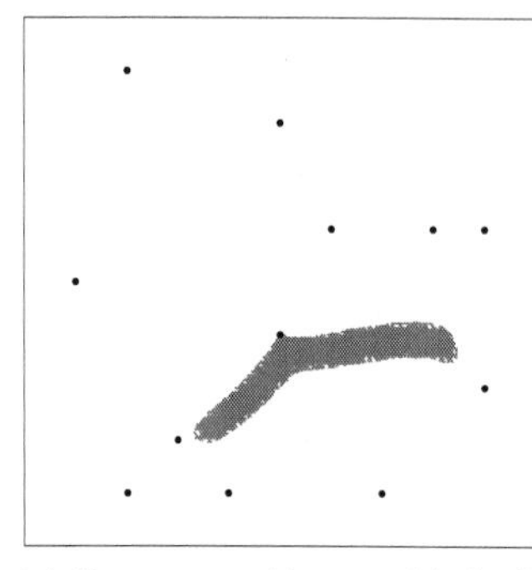

(a) Output set without multicriteria discarding tests.

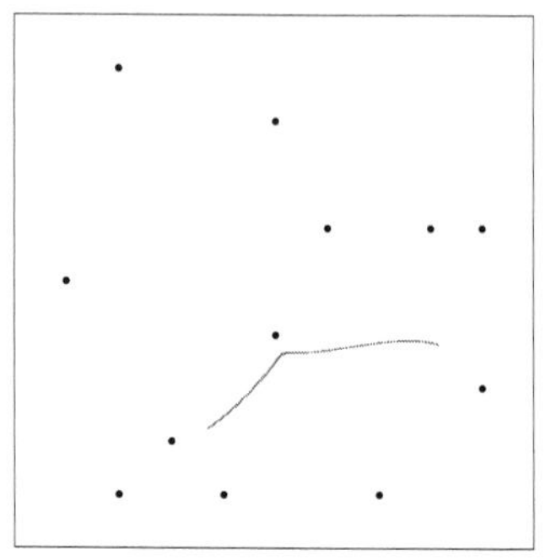

(b) Output set with multicriteria discarding tests.

Fig. 5.4 Results for Example 5.3. (a) Output set after 10,000 iterations without multicriteria discarding tests. (b) Output set after 10,000 iterations using the further discarding tests derived from Lemmas 5.1 and 5.2.

again lead to very sharp outer approximations of the set of all Pareto optimal solutions X_E.

Chapter 6
Extension for mixed combinatorial problems

Abstract All techniques presented in the previous chapters are dealing with pure continuous objective functions, therefore we now extend the geometric branch-and-bound algorithm to mixed continuous and combinatorial optimization problems. We are not only taking continuous variables into account but also some combinatorial variables. The extended algorithm is suggested in Section 6.1 and the rate of convergence again leads to a general convergence theory as shown in Section 6.2. Next, in Section 6.3 we derive some mixed bounding operations using techniques already discussed in Chapter 3 for pure continuous problems. Moreover, we show in Section 6.4 how to find exact optimal solutions under certain conditions which is demonstrated on some facility location problems in Section 6.5. Finally, we conclude the chapter with some numerical results in Section 6.6.

6.1 The algorithm

The technique suggested in this section is a generalization of the geometric branch-and-bound prototype algorithm presented in Chapter 2. Here, we extend the method to mixed integer minimization problems; that is to problems containing continuous and combinatorial variables. Hence, our goal is to minimize a function

$$f : \mathbb{R}^n \times \mathbb{Z}^m \to \mathbb{R}.$$

To this end, we assume a feasible area $X \times \Pi$, where $X \subset \mathbb{R}^n$ is a box and $\Pi \subset \mathbb{Z}^m$ with $|\Pi| < \infty$. To apply the algorithm, we need the following definition.

Definition 6.1. Let $X \subset \mathbb{R}^n$ be a box, $\Pi \subset \mathbb{Z}^m$ with $|\Pi| < \infty$, and consider

$$f : X \times \Pi \to \mathbb{R}.$$

A ***mixed bounding operation*** is a procedure to calculate for any subbox $Y \subset X$ a ***lower bound*** $LB(Y) \in \mathbb{R}$ with

$$LB(Y) \leq f(x,\pi) \text{ for all } x \in Y \text{ and } \pi \in \Pi$$

and to specify a point $r(Y) \in Y$ and a point $\kappa(Y) \in \Pi$.

Formally, we obtain the mixed bounding operation

$$(LB(Y), r(Y), \kappa(Y))$$

for all subboxes $Y \subset X$.

If a mixed bounding operation with a rate of convergence of at least $p = 1$ is known, see Sections 6.2 and 6.3, then the following algorithm finds a global minimum of

$$\min_{\substack{x \in X \\ \pi \in \Pi}} f(x,\pi)$$

up to an absolute accuracy of $\varepsilon > 0$.

1. Let $\mathscr{X}$ be a list of boxes and initialize $\mathscr{X} := \{X\}$.
2. Apply the mixed bounding operation to X and set $UB := f(r(X), \kappa(X))$.
3. If $\mathscr{X} = \emptyset$, the algorithm stops. Else set
$$\delta_{\max} := \max\{\delta(Y) : Y \in \mathscr{X}\}.$$
4. Select a box $Y \in \mathscr{X}$ with $\delta(Y) = \delta_{\max}$ and split it into s subboxes Y_1 to Y_s such that $Y = Y_1 \cup \cdots \cup Y_s$; see Subsection 2.3.2.
5. Set $\mathscr{X} = (\mathscr{X} \setminus Y) \cup \{Y_1, \ldots, Y_s\}$; that is delete Y from $\mathscr{X}$ and add Y_1 to Y_s.
6. Apply the mixed bounding operation to Y_1 to Y_s and set
$$UB = \min\{UB, f(r(Y_1), \kappa(Y_1)), \ldots, f(r(Y_s), \kappa(Y_s))\}.$$
7. For all $Z \in \mathscr{X}$, if $LB(Z) + \varepsilon \geq UB$ set $\mathscr{X} = \mathscr{X} \setminus Z$. If UB has not changed it is sufficient to check only the subboxes Y_1 to Y_s.
8. Whenever possible, apply some further discarding tests; see Section 6.4.
9. Return to Step 3.

6.2 Convergence theory

In order to evaluate the quality of bounding operations, we extend the definition for the rate of convergence as introduced in Section 2.4.

Definition 6.2. Let $X \subset \mathbb{R}^n$ be a box, let $\Pi \subset \mathbb{Z}^m$ with $|\Pi| < \infty$, and $f : X \times \Pi \to \mathbb{R}$. Furthermore, consider the minimization problem

$$\min_{\substack{x \in X \\ \pi \in \Pi}} f(x,\pi).$$

We say a mixed bounding operation has the ***rate of convergence*** $p \in \mathbb{N}$ if there exists a fixed constant $C > 0$ such that

$$f(r(Y), \kappa(Y)) - LB(Y) \leq C \cdot \delta(Y)^p \tag{6.1}$$

for all boxes $Y \subset X$.

The next theorem shows that the proposed algorithm terminates after a finite number of iterations if the mixed bounding operation has a rate of convergence of at least $p = 1$.

Theorem 6.1. *Let $X \subset \mathbb{R}^n$ be a box, let $\Pi \subset \mathbb{Z}^m$ with $|\Pi| < \infty$, and $f : X \times \Pi \to \mathbb{R}$. Furthermore, consider the minimization problem*

$$\min_{\substack{x \in X \\ \pi \in \Pi}} f(x, \pi)$$

and assume a mixed bounding operation with a rate of convergence of $p \geq 1$. Then for any $\varepsilon > 0$ the algorithm stops after finitely many steps with an ε-optimal solution.

Proof. Define the objective function

$$g(x) := \min_{\pi \in \Pi} f(x, \pi)$$

and consider the global optimization problem

$$\min_{x \in X} g(x).$$

Note that this problem is equivalent to the minimization of $f(x, \pi)$ on $X \times \Pi$.

Moreover, because $f(r(Y), \kappa(Y)) - LB(Y) \leq C \cdot \delta(Y)^p$ for any subbox $Y \subset X$, we obtain

$$\begin{aligned} g(r(Y)) - LB(Y) &= \min_{\pi \in \Pi} f(r(Y), \pi) - LB(Y) \\ &\leq f(r(Y), \kappa(Y)) - LB(Y) \leq C \cdot \delta(Y)^p. \end{aligned}$$

This yields a bounding operation for $g : X \to \mathbb{R}$ with a rate of convergence of $p \geq 1$. Hence, the theorem follows directly from Corollary 2.1. □

6.3 Mixed bounding operations

Our main goal in this section is to derive some mixed bounding operations with a rate of convergence of $p \geq 1$. Therefore, we assume that some combinatorial problems can be easily solved.

6.3.1 Mixed concave bounding operation

For our first bounding operation, assume that for any fixed $x \in X$ we are in a position to solve the combinatorial problem

$$\min_{\pi \in \Pi} f(x, \pi)$$

and furthermore assume that for any $\pi \in \Pi$ the function

$$f_\pi(x) := f(x, \pi)$$

is concave on X. Then, define

$$g(x) := \min_{\pi \in \Pi} f_\pi(x) = \min_{\pi \in \Pi} f(x, \pi)$$

and note that g is concave because the minimum of a finite number of concave functions is concave again. Denoting by $V(Y)$ the 2^n vertices of Y, we then obtain the ***mixed concave bounding operation***

$$\begin{aligned} LB(Y) &:= \min_{x \in Y} g(x) = \min_{v \in V(Y)} g(v), \\ r(Y) &\in \arg \min_{v \in V(Y)} g(v), \\ \kappa(Y) &\in \arg \min_{\pi \in \Pi} f(r(Y), \pi). \end{aligned}$$

Formally, we obtain the following result.

Theorem 6.2. *Consider $f : X \times \Pi \to \mathbb{R}$ and assume that for any fixed $\pi \in \Pi$ the function $f_\pi(x) = f(x, \pi)$ is concave.*

Then the mixed concave bounding operation has a rate of convergence of $p = \infty$.

Proof. The result is trivial because

$$f(r(Y), \kappa(Y)) = \min_{\substack{x \in Y \\ \pi \in \Pi}} f(x, \pi) = \min_{x \in Y} g(x) = LB(Y)$$

and we therefore find

$$f(r(Y), \kappa(Y)) - LB(Y) = 0$$

for all subboxes $Y \subset X$. □

6.3.2 Mixed d.c. bounding operation

The idea of the mixed d.c. bounding operation is to reduce the problem to the concave bounding operation as follows; see also Section 3.3.

We assume that for all $\pi \in \Pi$ a d.c. decomposition of $f_\pi(x) = f(x,\pi)$ can be constructed:

$$f_\pi(x) = g_\pi(x) - h_\pi(x),$$

where g_π and h_π are convex functions. Consider an arbitrary subbox $Y \subset X$ and $c = c(Y)$. Then, for any subgradient $\xi_\pi \in \mathbb{R}^n$ of g_π at c we obtain

$$a_\pi(x) := g_\pi(c) + \xi_\pi^T \cdot (x-c) - h_\pi(x) \le f_\pi(x) \text{ for all } x \in Y$$

and a_π is concave. Hence,

$$q(x) := \min_{\pi \in \Pi} a_\pi(x) \le \min_{\pi \in \Pi} f_\pi(x) = f(x,\pi) \text{ for all } x \in Y,\ \pi \in \Pi$$

and q is also concave. We now apply the concave bounding operation on q and obtain the ***mixed d.c. bounding operation***

$$\begin{aligned} LB(Y) &:= \min_{v \in V(Y)} q(v), \\ r(Y) &\in \arg\min_{v \in V(Y)} q(v), \\ \kappa(Y) &\in \arg\min_{\pi \in \Pi} a_\pi(r(Y)), \end{aligned}$$

where $V(Y)$ is again the set of the 2^n vertices of Y. Note that

$$LB(Y) = \min_{\pi \in \Pi} a_\pi(r(Y)) = a_{\kappa(Y)}(r(Y)).$$

Theorem 6.3. *Consider $f : X \times \Pi \to \mathbb{R}$ such that for all $\pi \in \Pi$ a d.c. decomposition of*

$$f_\pi(x) = g_\pi(x) - h_\pi(x)$$

is known and assume that g_π is twice continuously differentiable on X for all $\pi \in \Pi$.

Then the mixed d.c. bounding operation has a rate of convergence of $p = 2$.

Proof. For any $\pi \in \Pi$ and any subbox $Y \subset X$, we find

$$f_\pi(x) - a_\pi(x) \le C_\pi \cdot \delta(Y)^2$$

for some $C_\pi > 0$ that do not depend on Y; see Lemma 3.1. Defining

$$C_{\max} := \max\{C_\pi : \pi \in \Pi\},$$

we obtain

$$\begin{aligned} f(r(Y),\kappa(Y)) - LB(Y) &= f_{\kappa(Y)}(r(Y)) - a_{\kappa(Y)}(r(Y)) \\ &\le C_{\kappa(Y)} \cdot \delta(Y)^2 \le C_{\max} \cdot \delta(Y)^2 \end{aligned}$$

which proves the theorem. □

6.3.3 Mixed location bounding operation

Similar to the location bounding operation, see Section 3.9, assume that the objective function can be written as

$$f(x,\pi) = h(\pi, d(a_1,x),\ldots,d(a_m,x)),$$

where $a_1,\ldots,a_m \in \mathbb{R}^n$ are some given demand points and $d(a,x)$ is a given distance function. Furthermore, we assume that we can solve problems of the form

$$\min\{h(\pi,z) \ : \ \pi \in \Pi \text{ and } \ell_k \leq z_k \leq u_k \text{ for } k = 1,\ldots,m\},$$

where $\ell_k, u_k \in \mathbb{R}$ are some given constants for $k = 1,\ldots,m$. In order to calculate a lower bound $LB(Y)$ for an arbitrary box $Y \subset X$, suppose that the values

$$d_k^{\min}(Y) = \min\{d(a_k,x) \ : \ x \in Y\},$$
$$d_k^{\max}(Y) = \max\{d(a_k,x) \ : \ x \in Y\}$$

for $k = 1,\ldots,m$ are easily derived. This is the case if d is a monotone norm or a polyhedral gauge; see Plastria (1992). We then have the ***mixed location bounding operation***

$$LB(Y) := \min\{h(\pi,z) \ : \ \pi \in \Pi \text{ and } d_k^{\min}(Y) \leq z_k \leq d_k^{\max}(Y) \text{ for } k = 1,\ldots,m\},$$
$$r(Y) = c(Y),$$
$$\kappa(Y) \in \arg\min_{\pi\in\Pi} f(c(Y),\pi).$$

Theorem 6.4. *Assume that*

$$f(x,\pi) = h(\pi, d(a_1,x),\ldots,d(a_m,x)),$$

where $h : \Pi \times \mathbb{R}^m \to \mathbb{R}$ is for all $\pi \in \Pi$ a Lipschitzian function in the last m variables with constant $L_\pi \leq L_{\max}$ and assume that d is a norm.

Then the mixed bounding operation for location problems has a rate of convergence of $p = 1$.

Proof. Using the function

$$g(x) = \min_{\pi\in\Pi} f(x,\pi)$$

again, the proof is very similar to the result in Plastria (1992) and therefore it is omitted here. □

6.4 An exact solution method

For a given mixed integer optimization problem, we now want to extend the algorithm suggested in Section 6.1 in such a way that we obtain an exact global minimum. This is done by the following further discarding test; see Step 8 in Section 6.1.

We assume that for any fixed $\pi \in \Pi$ we are in a position to solve the pure continuous problem

$$\min_{x \in X} f(x, \pi).$$

Definition 6.3. Let $Y \subset X$ be a subbox of X. Then a set $\Omega(Y) \subset \Pi$ is called a ***combinatorial dominating set*** for Y if

$$\min_{\substack{x \in Y \\ \pi \in \Omega(Y)}} f(x, \pi) = \min_{\substack{x \in Y \\ \pi \in \Pi}} f(x, \pi).$$

In other words, if we know that $x \in Y$, we only have to consider $\pi \in \Omega(Y)$ and can neglect all $\pi \in \Pi \setminus \Omega(Y)$. Examples for sets $\Omega(Y)$ are given in the next section for some example problems.

If a combinatorial dominating set is known for every subbox $Y \subset X$, we suggest using the following further discarding test for a given parameter $M \in \mathbb{N}$.

8. For $i = 1, \ldots, s$, if $|\Omega(Y_i)| \leq M$ set $\mathscr{X} = \mathscr{X} \setminus Y_i$ and, furthermore, for all $\pi \in \Omega(Y_i)$ solve the minimization problem

$$t_{i,\pi} = \min_{x \in X} f(x, \pi)$$

and set $UB = \min\{UB, t_{i,\pi}\}$.

Note that the choice of the parameter $M \in \mathbb{N}$ strongly depends on the given objective function. If M is too small, the condition $|\Omega(Y_i)| \leq M$ might be satisfied only rarely. However, if M is too large, it could be too expensive to solve all $|\Omega(Y_i)|$ minimization problems. Therefore, we suggest starting the algorithm with a small value for M and increasing M throughout the algorithm if the method does not terminate.

Moreover, if we use $\varepsilon = 0$, we only delete subboxes $Z \in \mathscr{X}$ with $LB(Z) \geq UB$ in Step 7 of the prototype algorithm. Thus, we find an exact global minimum as long as the pure continuous problems can be solved exactly. Formally, we obtain the following result.

Lemma 6.1. *Let $X \subset \mathbb{R}^n$ be a box, let $\Pi \subset \mathbb{Z}^m$ with $|\Pi| < \infty$, and $f : X \times \Pi \to \mathbb{R}$. Moreover, assume that for any fixed $\pi \in \Pi$ the pure continuous problem*

$$\min_{x \in X} f(x, \pi)$$

can be solved exactly and consider an $M \geq 1$. Finally, assume that there exists a fixed constant $\tau > 0$ such that

$$|\Omega(Y)| \leq M \tag{6.2}$$

for all boxes $Y \subset X$ with $\delta(Y) \leq \tau$.

Then the geometric branch-and-bound algorithm for mixed combinatorial problems using $\varepsilon = 0$ and the previously presented discarding test finds an exact optimal solution for

$$\min_{\substack{x \in X \\ \pi \in \Pi}} f(x, \pi).$$

Proof. In every iteration of the algorithm a box with largest diameter is selected for a split into some smaller subboxes, therefore we obviously obtain

$$\delta(Y) \leq \tau$$

for all $Y \in \mathscr{X}$ after a finite number of iterations. Thus, the previously presented discarding test ensures the termination of the algorithm after a finite number of iterations.

Moreover, because $\varepsilon = 0$, no box Y that contains an optimal solution will be discarded throughout the algorithm. Hence, the lemma is shown because the pure continuous problems can be solved exactly. □

Note that Lemma 6.1 holds independently from the bounding operation.

Unfortunately, the condition given in Equation (6.2) is in general not satisfied for any $M < |\Pi|$ in our following example problems; see Example 6.2. However, with randomly generated input data we found an exact optimal solution in all of our numerical studies; see Section 6.6.

6.5 Example problems

In this section, two example problems are discussed, namely the truncated Weber problem and the multisource Weber problem. For both problems, consider m demand points $a_1, \ldots, a_m \in \mathbb{R}^2$.

6.5.1 The truncated Weber problem

The ***truncated Weber problem*** is to find a new facility location $x \in \mathbb{R}^2$ such that the sum of the weighted distances between the new facility location x and the nearest $1 < K < m$ demand points is minimized. Therefore, assume some weights $w_1, \ldots, w_m \geq 0$, let $X \subset \mathbb{R}^2$ be a box, and consider

$$\Pi = \{\pi = (\pi_1, \ldots, \pi_m) \in \{0,1\}^m \; : \; \pi_1 + \cdots + \pi_m = K\}.$$

Thus, the truncated Weber problem is to minimize the objective function

$$f(x,\pi) = \sum_{k=1}^{m} \pi_k \cdot w_k \cdot d(a_k,x) \quad \text{for } x \in X,\ \pi \in \Pi,$$

where d is a given distance function. Note that this problem is a special case of the general ordered median problem, see Nickel and Puerto (2005), and moreover a special case of the general location-allocation problem introduced by Plastria and Elosmani (2008). Solution algorithms can be found in Drezner and Nickel (2009a,b). But note that these methods do not provide an exact optimal solution.

Obviously, for any fixed $x \in X$ we can easily solve the problem

$$\min_{\pi \in \Pi} f(x,\pi)$$

by sorting $w_k \cdot d(a_k,x)$ for $k = 1,\ldots,m$. Moreover, for any fixed $\pi \in \Pi$ the functions $f_\pi(x) = f(x,\pi)$ are classical Weber problem objective functions and therefore convex. Hence, we can apply the d.c. bounding operation.

Next, denote by

$$\begin{aligned} d_k^{\min}(Y) &= \min\{w_k \cdot d(a_k,x) \ : \ x \in Y\}, \\ d_k^{\max}(Y) &= \max\{w_k \cdot d(a_k,x) \ : \ x \in Y\} \end{aligned}$$

the minimal and the maximal weighted distances between any subbox Y and the demand point a_k, respectively. Defining

$$\Omega_k(Y) := \begin{cases} \{0\} & \text{if } d_k^{\min}(Y) > d_j^{\max}(Y) \text{ for at least } K \text{ indices } j \in \{1,\ldots,m\} \\ \{1\} & \text{if } d_k^{\max}(Y) > d_j^{\min}(Y) \text{ for at most } K \text{ indices } j \in \{1,\ldots,m\} \\ \{0,1\} & \text{else} \end{cases}$$

for $k = 1,\ldots,m$, we find

$$\Omega(Y) := \Pi \cap (\Omega_1(Y) \times \cdots \times \Omega_m(Y)).$$

Example 6.1. As an example with $m = 7$ consider the values of $d_k^{\min}(Y)$ and $d_k^{\max}(Y)$ for $k = 1,\ldots,7$ as depicted in Figure 6.1.

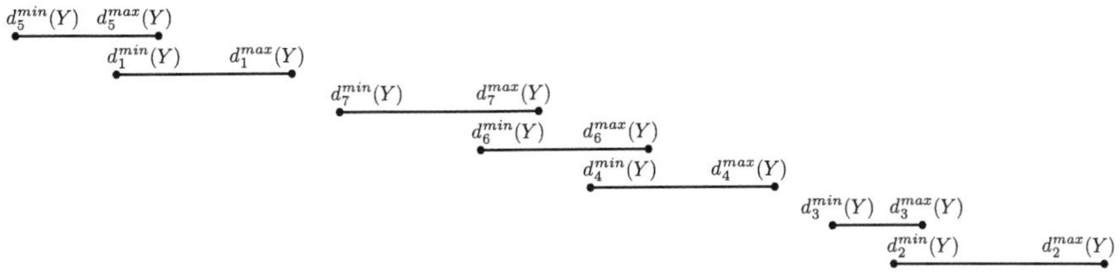

Fig. 6.1 Example for the calculation of $\Omega(Y)$ for the truncated Weber problem.

For the case $K = 3$, we obtain the following sets.

$$\begin{array}{llll} \Omega_1 = \{1\}, & \Omega_2 = \{0\}, & \Omega_3 = \{0\}, & \Omega_4 = \{0\}, \\ \Omega_5 = \{1\}, & \Omega_6 = \{0,1\}, & \Omega_7 = \{0,1\}. & \end{array}$$

Hence, for the given box Y only the variables π_6 and π_7 are not assigned to a fixed value. Furthermore, note that $|\Omega(Y)| = 2$.

The definition of $\Omega(Y)$ yields the following result.

Lemma 6.2. *Consider the truncated Weber problem with $f(x,\pi)$, let $Y \subset X$ be an arbitrary subbox, and assume that $(x^*, \pi^*) \in Y \times \Pi$ is an optimal solution for*

$$\min_{\substack{x \in Y \\ \pi \in \Pi}} f(x, \pi).$$

Then we have $\pi^* \in \Omega(Y)$; that is $\Omega(Y)$ is a combinatorial dominating set for Y.

Proof. Assume that $\pi^* \notin \Omega(Y)$. We now construct for all $x \in Y$ a $\mu = (\mu_1, \ldots, \mu_m) \in \Omega(Y)$ such that $f(x, \pi^*) > f(x, \mu)$.

To this end, let $I \subset \{1, \ldots, m\}$ such that $i \in I$ if and only if $\pi_i^* \notin \Omega_i(Y)$. Next, define

$$v_i = \begin{cases} 0 \text{ if } \pi_i^* = 1, \\ 1 \text{ if } \pi_i^* = 0 \end{cases}$$

for all $i \in I$ and find a $\mu \in \Omega(Y)$ such that $\mu_i = v_i$ for all $i \in I$. Note that such a μ exists and note that

$$\sum_{\substack{k=1 \\ \mu_k \neq \pi_k^*}}^{m} \pi_k^* = \sum_{\substack{k=1 \\ \mu_k \neq \pi_k^*}}^{m} \mu_k.$$

Moreover, by construction of $\Omega(Y)$ we obtain for all $x \in Y$,

$$\sum_{\substack{k=1 \\ \mu_k \neq \pi_k^*}}^{m} \pi_k^* \cdot w_k \cdot d(a_k, x) > \sum_{\substack{k=1 \\ \mu_k \neq \pi_k^*}}^{m} \mu_k \cdot w_k \cdot d(a_k, x);$$

see also Example 6.1. Hence, for all $x \in Y$ we have

$$f(x, \pi^*) > f(x, \mu),$$

a contradiction to the optimality of (x^*, π^*). □

Unfortunately, in general we cannot expect

$$|\Omega(Y)| \leq M$$

for all boxes $Y \subset X$ with $\delta(Y) \leq \tau$ for a $\tau \geq 0$ and an $M < |\Pi|$ as the following counterexample shows.

Example 6.2. Consider the m demand points $a_k = (k/m, 1-k/m) \in \mathbb{R}^2$ for $k = 1,\ldots,m$ and consider the objective function

$$f(x,\pi) = \sum_{k=1}^{m} \pi_k \cdot \|x - a_k\|_1.$$

Then, for all $1 < K < m$ and $Y = [0,0] \times [0,0]$ we have

$$d_k^{\min}(Y) = d_k^{\max}(Y) = 1 \text{ for } k = 1,\ldots,m.$$

Hence, $\Omega(Y) = \Pi$ although $\delta(Y) = 0$.

6.5.2 The multisource Weber problem

The ***multisource Weber problem*** is to locate p new facilities $x_1,\ldots,x_p \in \mathbb{R}^2$ assuming that each demand point is served by its nearest new facility. To be more precise, with $X \subset \mathbb{R}^{2p}$, with $P = \{1,\ldots,p\}$, and with $\Pi = P^m \subset \mathbb{Z}^m$ the problem is to minimize the objective function

$$f(x,\pi) = f(x_1,\ldots,x_p,\pi_1,\ldots,\pi_m) = \sum_{k=1}^{m} w_k \cdot d(a_k, x_{\pi_k}),$$

where $w_1,\ldots,w_m \geq 0$ are some given nonnegative weights.

The multisource Weber problem is one of the most studied facility location problems. For example, Drezner (1984) presented an exact algorithm for $p = 2$ using the Euclidean norm and some more general global optimization approaches can be found in Chen et al. (1998) and Schöbel and Scholz (2010a). Furthermore, the variable neighborhood search heuristic was applied in Brimberg et al. (2004) and Brimberg et al. (2006).

Here, we can make use of the d.c. bounding operation again because the objective function is convex for all fixed $\pi \in \Pi$.

Next, let $Y = Y_1 \times \cdots \times Y_p$ such that $x_i \in Y_i \subset \mathbb{R}^2$ for the new facilities $i = 1,\ldots,p$ and define the sets $\Omega_k(Y) \subset P$ for $k = 1,\ldots,m$ as follows.

$$\Omega_k(Y) := P \setminus \Big\{ i \in P : \text{ there is a } j \in P \text{ with } d_k^{\min}(Y_i) > d_k^{\max}(Y_j) \Big\},$$

where $d_k^{\min}(Y_i)$ and $d_k^{\max}(Y_i)$ are defined as before; see also Example 6.3.

Example 6.3. As an example with $p = 7$, consider the values of $d_1^{\min}(Y_i)$ and $d_1^{\max}(Y_i)$ for $i = 1,\ldots,7$ as depicted in Figure 6.2.

In this case, we obtain $\Omega_1(Y) = \{2,3,6\}$. Thus, for all $x = (x_1,\ldots,x_7) \in Y$ the demand point a_1 will never be served by x_1, x_4, x_5, or x_7.

This leads to

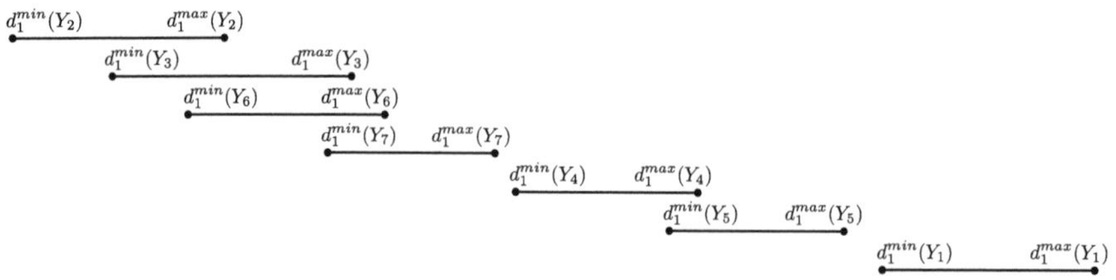

$d_1^{min}(Y_2)$ $d_1^{max}(Y_2)$
$d_1^{min}(Y_3)$ $d_1^{max}(Y_3)$
$d_1^{min}(Y_6)$ $d_1^{max}(Y_6)$
$d_1^{min}(Y_7)$ $d_1^{max}(Y_7)$
$d_1^{min}(Y_4)$ $d_1^{max}(Y_4)$
$d_1^{min}(Y_5)$ $d_1^{max}(Y_5)$
$d_1^{min}(Y_1)$ $d_1^{max}(Y_1)$

Fig. 6.2 Example for the calculation of $\Omega_1(Y)$ for the multisource Weber problem.

$$\Omega(Y) \;:=\; \Omega_1(Y) \times \cdots \times \Omega_m(Y)$$

and by construction of $\Omega(Y)$ we obtain the following result.

Lemma 6.3. *Consider the multisource Weber problem with $f(x,\pi)$, let $Y \subset X$ be an arbitrary subbox, and assume that $(x^*,\pi^*) \in Y \times \Pi$ is an optimal solution for*

$$\min_{\substack{x\in Y\\ \pi\in\Pi}} f(x,\pi).$$

Then we have $\pi^* \in \Omega(Y)$; that is $\Omega(Y)$ is a combinatorial dominating set for Y.

Proof. Assume that $\pi^* \notin \Omega(Y)$. Hence, our goal is again to construct for all $x \in Y$ a $\mu = (\mu_1,\ldots,\mu_m) \in \Omega(Y)$ such that $f(x,\pi^*) > f(x,\mu)$.

To this end, assign $\mu_k = \pi_k^*$ for all $k \in \{1,\ldots,m\}$ with $\pi_k^* \in \Omega_k(Y)$:

$$d(a_k, x_{\pi_k^*}) \;=\; d(a_k, x_{\mu_k})$$

for all $x = (x_1,\ldots,x_p) \in Y$. Moreover, note that for all $x = (x_1,\ldots,x_p) \in Y$ and all $k \in \{1,\ldots,m\}$ with $\pi_k^* \notin \Omega_k(Y)$ there exists by construction of $\Omega_k(Y)$ a $\mu_k \in \Omega_k(Y)$ such that

$$d(a_k, x_{\pi_k^*}) \;>\; d(a_k, x_{\mu_k});$$

see also Example 6.3. To sum up, for all $x \in Y$ there exists a $\mu \in \Omega(Y)$ such that

$$f(x,\pi^*) \;>\; f(x,\mu),$$

a contradiction to the optimality of (x^*,π^*). □

6.6 Numerical results

For our example problems, we generated up to $m = 10{,}000$ demand points $a_1,\ldots,a_m$ uniformly distributed in $X = \{-500,-499,\ldots,499,500\}^2$ and positive weights $w_k \in \{1,2,\ldots,10\}$ for $k = 1,\ldots,m$. Ten problems were run for different values of m. As a distance measure, we applied the rectilinear norm:

$$d(a_k,x) = \|x-a_k\|_1$$

for $k=1,\ldots,m$. Hence, the pure facility location problems could be solved easily up to an exact optimal solution using the algorithm given in Section 1.3 with a complexity of $\mathcal{O}(m\cdot\log m)$; see also Drezner et al. (2001). Moreover, we used $\varepsilon=0$ and $M=4$ throughout the algorithm and note again that we found an exact optimal solution in all problem instances.

Furthermore, we remark that using the rectilinear norm, both problems can also be solved by a finite dominating set: one can construct a finite set of solutions that contains at least one global minimum.

6.6.1 The truncated Weber problem

In our first example problem, we set $K=m/5$ and all boxes were split into $s=4$ congruent subboxes. Our results are illustrated in Table 6.1. Therein, the runtimes, the number of iterations throughout the branch-and-bound algorithm, and the number of solved single facility location problems are reported.

	Runtime (Sec.)			Iterations			Location Problems		
m	Min	Max	Ave.	Min	Max	Ave.	Min	Max	Ave.
10	0.00	0.16	0.05	24	47	32.2	10	42	23.3
20	0.01	0.15	0.04	34	52	43.9	15	51	26.7
50	0.01	0.30	0.05	44	72	52.2	8	30	19.0
100	0.04	0.53	0.10	54	95	73.9	14	65	28.4
200	0.10	0.42	0.16	67	137	96.1	6	62	23.5
500	0.30	0.91	0.44	87	186	112.5	6	25	14.7
1,000	0.78	1.37	1.07	106	178	140.4	12	38	21.4
2,000	2.37	5.47	3.04	145	331	183.8	8	122	31.0
5,000	6.14	15.93	9.43	142	360	215.9	11	182	33.6
10,000	20.03	32.35	27.38	205	338	284.3	2	139	38.8

Table 6.1 Numerical results for the truncated Weber problem.

Observe that the average number of solved facility location problems is for all values of m almost constant.

6.6.2 The multisource Weber problem

The multisource Weber problem was solved for $p=2$ and $p=3$. Here, every selected box Y throughout the algorithm was split into $s=2$ subboxes; that is Y was bisected in two subboxes perpendicular to the direction of the maximum width component. Our results are collected in Table 6.2.

		Runtime (Sec.)			Iterations			Location Problems		
m	p	Min	Max	Ave.	Min	Max	Ave.	Min	Max	Ave.
10	2	0.01	0.17	0.05	223	788	463.1	11	261	115.4
20	2	0.02	0.36	0.08	319	1,728	926.1	48	501	200.3
50	2	0.08	0.38	0.23	637	2,450	1,463.5	9	379	119.4
100	2	0.21	1.11	0.50	779	2,700	1,688.6	2	472	126.7
200	2	1.06	2.77	1.79	1,929	4,890	3,204.0	44	1,451	258.4
500	2	2.81	7.50	4.52	2,152	5,185	3,385.4	13	303	139.0
1,000	2	3.92	20.59	12.60	1,401	7,611	4,747.1	2	395	128.8
2,000	2	13.30	59.58	34.20	2,568	11,015	6,392.6	2	979	247.2
5,000	2	49.35	277.35	128.60	3,700	20,085	9,450.5	2	2,711	423.1
10	3	0.46	1.66	0.96	3,409	11,707	6,734.0	286	2,182	947.9
20	3	1.88	26.98	6.61	7,271	79,666	22,602.1	567	7,663	2,879.6
50	3	9.38	79.60	33.42	11,605	95,584	45,969.4	152	7,333	2,660.9
100	3	41.56	217.94	128.50	32,710	133,468	83,211.5	112	10,028	3,079.1

Table 6.2 Numerical results for the multisource Weber problem.

As can be seen, for the multisource Weber problem the number of solved pure location problems increases with the number m of demand points. However, even with $m = 5,000$ the multisource Weber problem with two new facilities could be solved taking only a few hundred single facility problems into account.

Chapter 7
The circle detection problem

Abstract In the previous chapters we discussed the theory of geometric branch-and-bound methods and several extensions were given. In this chapter, we now present an application of the suggested techniques. To be more precise, we show how global optimization can be used in image processing to detect imperfect instances of shapes such as lines, circles, or ellipses. We introduce the problem in Section 7.1 and some useful notations are collected in Section 7.2. Before the general problem can be formulated, we have to detect edges in images. To this end, Canny's edge detection algorithm is briefly summarized in Section 7.3. Next, we can formulate the detection problem as a global optimization problem in Section 7.4. In the following, we also discuss the circle detection problem in detail. Some lower bounds are suggested in Section 7.5 before Section 7.6 illustrates the proposed circle detection problem on several examples.

7.1 Introduction

In this chapter, we describe the use of global optimization techniques in image processing: for a given image we show how to detect imperfect pictured shapes such as lines, circles, and ellipses. In almost all commonly used detection algorithms, we first of all have to detect edges in the given image; see Section 7.3 or Jähne (2002). Once the edges are found, different approaches for detecting shapes can be found in the literature.

For example, the ***Hough transform*** projects every detected edge point to a parameter space where the parameters represent the shape to be found, for instance the center and the radius for a circle. Next, a voting procedure is carried out in the parameter space and object candidates are obtained as local maxima; see Duda and Hart (1972). Many extensions and speed-up techniques for the Hough transform can be located in the literature; see, for example, Yip et al. (1992) and Guil and Zapata (1997) for detecting circles and ellipses.

Another technique is the ***random sample consensus*** or ***Ransac*** method for short; see Fischler and Bolles (1981). The basic idea is to divide the set of detected edge points into ***outliers***, points that do not fit to the model, and ***inliers***, points that do fit to the model. This is done by a random selection of some data points evaluated on a fitness function. Applications of Ransac can be found in Clarke et al. (1996) as well as Rousseeuw and Leroy (1987) and references therein.

We remark that these methods also find applications in statistics because they are related to maximum likelihood estimators. For robust fitting of mixtures based on the trimmed likelihood estimator, see Neykov et al. (2007).

However, the Hough transform yields in general only local optima and Ransac finds a reasonable result only with a certain probability, therefore our goal is to formulate the detection problem as a global optimization problem. With the help of lower bounds related to the techniques presented in Chapter 3 we can find global optimal solutions. Global optimization approaches in image processing have also been used in Breuel (2003a,b). However, lower bounds therein were only calculated by interval analysis which turns out to be slow or inaccurate.

In this chapter we discuss the circle detection problem in detail. Moreover, it is shown that some other shapes such as lines or ellipses can be detected in the same manner.

7.2 Notations

Before the edge detection algorithm is presented in the next section, we introduce some notations to formulate the technique in a mathematically correct way.

Definition 7.1. A (grayscaled) ***image*** M is given by a function

$$M : \{1,\dots,w\} \times \{1,\dots,h\} \to \mathbb{R},$$

where w is the ***width*** and h is the ***height*** of the image M and $M(i,j)$ represents the intensity of the point (i,j).

Moreover, a function

$$F : \{-r,\dots,r\} \times \{-r,\dots,r\} \to \mathbb{R}$$

is called a ***filter***. To be more precise, we use the following notation.

Definition 7.2. The ***discrete convolution*** of an image

$$M : \{1,\dots,w\} \times \{1,\dots,h\} \to \mathbb{R}$$

and a filter

$$F : \{-r,\dots,r\} \times \{-r,\dots,r\} \to \mathbb{R}$$

is defined by $(M * F) : \{1,\dots,w\} \times \{1,\dots,h\} \to \mathbb{R}$ with

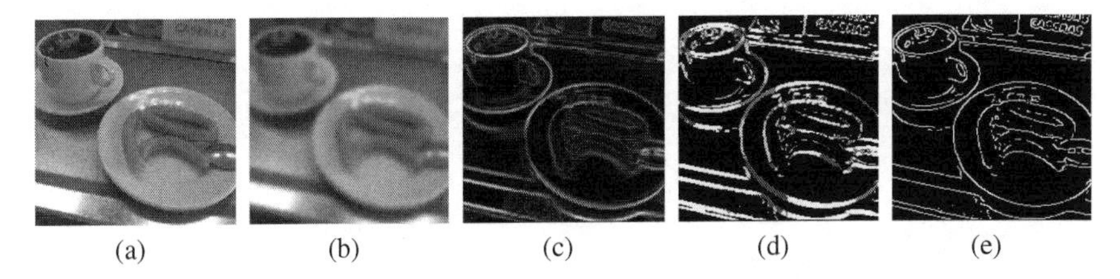

(a) (b) (c) (d) (e)

Fig. 7.1 Illustration of Canny's edge detection algorithm. (a) Original image M as input for the algorithm. (b) Image G after applying a Gaussian filter. (c) Edge intensity image I. (d) Four-color image R using angle rounding. (e) The detected edges given as the image B.

$$(M*F)(x,y) \;=\; \sum_{i=-r}^{r}\sum_{j=-r}^{r} M(x+i,y+j)\cdot F(i,j).$$

If $(x+i,y+j)\notin\{1,\dots,w\}\times\{1,\dots,h\}$ define $M(x+i,y+j):=M(x,y)$.

Note that the convolution $(M*F)$ is an image again. Moreover, a filter is represented in the following by a matrix $F=(f_{ij})\in\mathbb{R}^{(2r+1)\times(2r+1)}$:

$$F(i,j) \;=\; f_{i+r+1,\,j+r+1}.$$

Remark 7.1. Although an image is defined by a discrete function, the gradient and the edge of an image should be understood as follows.

Assume that an image M is defined by a differentiable function $M:X\to\mathbb{R}$, where $X\subset\mathbb{R}^2$ is a box. Hence, the gradient of M at $x\in X$ is simply given by $\nabla M(x)$. Furthermore, we say that $x\in X$ is an edge point of M if the absolute value of $\nabla M(x)$ is greater than a given threshold. In our discrete case, the idea of ***gradient*** and ***edge point*** in an image $M:\{1,\dots,w\}\times\{1,\dots,h\}\to\mathbb{R}$ is exactly the same except that we consider the numerical differentiation to calculate the gradient.

7.3 Canny edge detection

Using the notations given in the previous section, we now briefly discuss the ***Canny's edge detection*** technique we use to obtain the circle detection problem; see Canny (1986). All steps throughout the algorithm are illustrated on an example image in Figure 7.1.

As input for the algorithm consider an image

$$M:\{1,\dots,w\}\times\{1,\dots,h\}\to\{0,\dots,255\},$$

where $M(x,y)$ represents the intensity or greyscale of the pixel (x,y); see Figure 7.1(a). To be more precise, we define $M(x,y)=0$ if (x,y) is black, $M(x,y)=255$ if (x,y) is white, and anything in between represents the corresponding brightness of (x,y). The goal is to detect edges in M; see Remark 7.1.

Noise reduction.

In order to avoid some noise in the given image M, we apply a ***Gaussian filter***, for example,

$$F_\sigma = \frac{1}{37} \cdot \begin{pmatrix} 2 & 5 & 2 \\ 5 & 9 & 5 \\ 2 & 5 & 2 \end{pmatrix} \quad \text{or} \quad F_\sigma = \frac{1}{159} \cdot \begin{pmatrix} 2 & 4 & 5 & 4 & 2 \\ 4 & 9 & 12 & 9 & 4 \\ 5 & 12 & 15 & 12 & 5 \\ 4 & 9 & 12 & 9 & 4 \\ 2 & 4 & 5 & 4 & 2 \end{pmatrix}.$$

Note that these examples are a discretization of the normal distribution

$$g(x,y) = \frac{1}{2\pi\sigma} \cdot \exp\left(-\frac{1}{2\sigma}(x^2+y^2)\right)$$

with $\sigma = 1$ and $\sigma = 1.4$, respectively. Define the image $G := (M * F_\sigma)$, see Figure 7.1(b).

Gradient calculation.

In the next step, we calculate the gradient of G by applying an edge detection filter on G. Therefore consider, for example, the ***Prewitt filters***

$$F_x = \begin{pmatrix} -1 & 0 & 1 \\ -1 & 0 & 1 \\ -1 & 0 & 1 \end{pmatrix} \quad \text{and} \quad F_y = \begin{pmatrix} -1 & -1 & -1 \\ 0 & 0 & 0 \\ 1 & 1 & 1 \end{pmatrix}$$

or the ***Sobel filters***

$$F_x = \begin{pmatrix} 1 & 0 & -1 \\ 2 & 0 & -2 \\ 1 & 0 & -1 \end{pmatrix} \quad \text{and} \quad F_y = \begin{pmatrix} 1 & 2 & 1 \\ 0 & 0 & 0 \\ -1 & -2 & -1 \end{pmatrix}.$$

Then we define the images $G_x := (G * F_x)$ and $G_y := (G * F_y)$.

Edge intensity and angle finding.

The edge intensity and the edge angle are now given by the images

$$I, \Theta : \{1,\dots,w\} \times \{1,\dots,h\} \to \mathbb{R}$$

defined by

$$I(x,y) = \sqrt{(G_x(x,y))^2 + (G_y(x,y))^2},$$

$$\Theta(x,y) = \begin{cases} \frac{\pi}{2} & \text{if } G_y(x,y) = 0 \\ \arctan\left(\frac{G_x(x,y)}{G_y(x,y)}\right) & \text{if } G_y(x,y) \neq 0 \end{cases},$$

respectively; see Figure 7.1(c).

Angle rounding.

The angles are rounded such that we obtain an image

$$R : \{1,\dots,w\} \times \{1,\dots,h\} \to \{0,\dots,4\}.$$

Therefore, consider a given threshold γ, for example,

$$\gamma := \frac{1}{8} \cdot \max\{I(x,y) \ : \ 1 \le x \le w,\ 1 \le y \le h\}.$$

We then define

$$R(x,y) = \begin{cases} 0 & \text{if } I(x,y) \le \gamma \\ 1 & \text{if } I(x,y) > \gamma \text{ and } \Theta(x,y) \in \left(-\frac{\pi}{2}, -\frac{3\pi}{8}\right] \\ 2 & \text{if } I(x,y) > \gamma \text{ and } \Theta(x,y) \in \left(-\frac{3\pi}{8}, -\frac{\pi}{8}\right] \\ 3 & \text{if } I(x,y) > \gamma \text{ and } \Theta(x,y) \in \left(-\frac{\pi}{8}, \frac{\pi}{8}\right] \\ 4 & \text{if } I(x,y) > \gamma \text{ and } \Theta(x,y) \in \left(\frac{\pi}{8}, \frac{3\pi}{8}\right] \\ 1 & \text{if } I(x,y) > \gamma \text{ and } \Theta(x,y) \in \left(\frac{3\pi}{8}, \frac{\pi}{2}\right] \end{cases};$$

see Figure 7.1(d).

Nonmaximum suppression.

Next, the image R leads to the final binary image

$$B : \{1,\dots,w\} \times \{1,\dots,h\} \to \{0,1\}$$

defined as follows:

$$B(x,y) = \begin{cases} 1 & \text{if } R(x,y) = 1 \text{ and } I(x,y) > \max\{I(x-1,y), I(x+1,y)\} \\ 1 & \text{if } R(x,y) = 2 \text{ and } I(x,y) > \max\{I(x-1,y-1), I(x+1,y+1)\} \\ 1 & \text{if } R(x,y) = 3 \text{ and } I(x,y) > \max\{I(x,y-1), I(x,y+1)\} \\ 1 & \text{if } R(x,y) = 4 \text{ and } I(x,y) > \max\{I(x-1,y+1), I(x+1,y-1)\} \\ 0 & \text{else} \end{cases};$$

see Figure 7.1(e). If $(x+i, y+j) \notin \{1,\dots,w\} \times \{1,\dots,h\}$ for $i, j \in \{-1,0,1\}$ define $I(x+i, y+j) = 0$.

The output of Canny's edge detection technique is the image B which is called an ***edge image***. Note that if and only if $B(x,y) = 1$ the algorithm found out that the pixel (x,y) belongs to an edge. We remark that a more sophisticated edge detection method using second- and third-order derivatives computed from a scale-space representation can be found in Lindeberg (1998).

7.4 Problem formulation

Using our preliminary discussions, the goal in this section is a formulation of the detection problem for general shapes making use of global optimization. However, note that only the circle detection problem is discussed in the following sections.

Consider an image

$$M : \{1,\dots,w\} \times \{1,\dots,h\} \to \{0,\dots,255\},$$

where some shapes have to be detected and consider the corresponding edge image

$$B : \{1,\dots,w\} \times \{1,\dots,h\} \to \{0,1\}$$

found by Canny's edge detection algorithm. Moreover, define

$$\mathscr{A} := \{(x,y) \in \mathbb{R}^2 : x \in \{1,\dots,w\},\ y \in \{1,\dots,h\},\ B(x,y) = 1\}.$$

Then, our general objective function to be minimized is

$$f(x) = \sum_{a \in \mathscr{A}} \phi(d_a(x));$$

see Breuel (2003b). Here, $d_a : \mathbb{R}^n \to [0,\infty)$ is the Euclidean distance between $a \in \mathscr{A}$ and a shape given by the parameters $x = (x_1,\dots,x_n)$. Furthermore, $\phi : [0,\infty) \to (-\infty,0]$ is a nondecreasing ***weighting function***.

Some examples of weighting functions are

$$\phi(t) = \begin{cases} -1 \text{ if } t \leq \delta \\ \ \ 0 \text{ if } t > \delta \end{cases}, \quad \phi(t) = \min\left\{0, \frac{t^2}{\delta^2} - 1\right\}, \quad \phi(t) = -\exp\left(-\frac{1}{\delta} \cdot t^2\right)$$

for a given threshold δ; see, for instance, Breuel (2003b).

Moreover, because a circle is given by its center $x \in \mathbb{R}^2$ and radius $r \geq 0$, for the circle detection problem, for example, we obtain

$$d_a(x,r) = \Big| \|x-a\|_2 - r \Big|.$$

A similar approach for detecting shapes in images can be found in Breuel (2003a,b). Therein, the author solved geometric matching problems under translation and rotation which also yield a global optimization problem. The suggested solution approach is a geometric branch-and-bound method with lower bounds derived from interval analysis. But inasmuch as these bounds are not sharp enough, the algorithm is slow or inaccurate. Our goal in the next section is to present some more sophisticated lower bounds for the circle detection problem.

7.4.1 The circle detection problem

Although the previous discussion directly leads to the circle detection problem, we introduce a ***penalty term*** as follows. We expect more points $a \in \mathscr{A}$ close to the circumference for larger circles, thus a term proportional to the radius is added to the objective function. This penalty term is, in particular, suitable if some smaller circles are to be detected.

To sum up, the ***circle detection problem*** using the third weighting function and a penalty term with factor $C > 0$ is to minimize the objective function

$$f(x,r) = \sum_{a\in\mathscr{A}} \phi(d_a(x,r)) + C\cdot r = -\sum_{a\in\mathscr{A}} \exp\left(-\frac{1}{\delta}\cdot(\|x-a\|_2 - r)^2\right) + C\cdot r. \quad (7.1)$$

7.5 Bounding operation

For any box $Y \subset \mathbb{R}^2 \times [0,\infty)$ we find a simple lower bound on the objective function (7.1) using the natural interval bounding operation with a rate of convergence of $p = 1$. However, as mentioned before these bounds are, in particular, not tight enough for smaller boxes and our goal is to find some more sophisticated lower bounds.

To this end, for all $a \in \mathscr{A}$ define

$$q_a(x,r) := (\|x-a\|_2 - r)^2.$$

The next step is to find a d.c. decomposition of q_a. Because $\|x-a\|_2 \geq 0$ and $r \geq 0$, due to Lemma 1.10 we have

$$2\cdot(\|x-a\|_2\cdot r) = (\|x-a\|_2 + r)^2 - (\|x-a\|_2^2 + r^2).$$

Hence, we obtain

$$\begin{aligned} q_a(x,r) = (\|x-a\|_2 - r)^2 &= (\|x-a\|_2^2 + r^2) - 2\cdot(\|x-a\|_2\cdot r) \\ &= (\|x-a\|_2^2 + r^2) - (\|x-a\|_2 + r)^2 + (\|x-a\|_2^2 + r^2) \end{aligned}$$

$$= 2 \cdot \left(\|x-a\|_2^2 + r^2 \right) - \left(\|x-a\|_2 + r \right)^2 \;=\; g_a(x,r) - h_a(x,r),$$

where $g_a(x,r) = 2\left(\|x-a\|_2^2 + r^2\right)$ and $h_a(x,r) = \left(\|x-a\|_2 + r\right)^2$ are convex functions. This d.c. decomposition yields

$$m_a(x,r) \;:=\; g_a(c) + \xi_a^T \cdot ((x,r) - c) \;-\; h_a(x,r) \;\leq\; q_a(x,r) \tag{7.2}$$

for all $(x,r) \in Y$, where $c = c(Y) \in \mathbb{R}^3$ and ξ_a is a subgradient of g_a at c. The definition of m_a leads to the following bounding operation.

Theorem 7.1. *Define the function* $z : \mathbb{R}^3 \to \mathbb{R}$ *by*

$$z(x,r) \;:=\; -\sum_{a \in \mathscr{A}} \exp\left(-\frac{1}{\delta} \cdot m_a(x,r) \right) + C \cdot r.$$

Then a bounding operation for the circle detection problem is

$$LB(Y) \;:=\; \min_{v \in V(Y)} z(v) \;\text{ and }\; r(Y) \;\in\; \arg \min_{v \in V(Y)} z(v),$$

where $V(Y)$ *is the set of the eight vertices of* $Y \subset \mathbb{R}^2 \times [0,\infty)$.

Proof. We only have to show that $LB(Y) \leq f(x,r)$ for all $(x,r) \in Y$. To this end, consider the scalar function $\varphi(t) = -\exp(-t/\delta)$. Because φ is monotone increasing, we find

$$\varphi(m_a(x,r)) \;\leq\; \varphi(q_a(x,r)) \;=\; \varphi((\|x-a\|_2 - r)^2),$$

see Equation (7.2). Moreover, because m_a is concave and φ is not only monotone increasing but also concave, we obtain that $\varphi \circ m_a$ are concave functions for all $a \in \mathscr{A}$. Hence, the function z is concave and because

$$z(x,r) \;=\; \sum_{a \in \mathscr{A}} \varphi(m_a(x,r)) + C \cdot r \;\leq\; \sum_{a \in \mathscr{A}} \varphi((\|x-a\|_2 - r)^2) + C \cdot r \;=\; f(x,r),$$

it is shown that $LB(Y) \leq f(x,r)$ for all $(x,r) \in Y$. □

Furthermore, note that if f is differentiable at c, then $f(c) = z(c)$ and $\nabla f(c) = \nabla z(c)$. Hence, Lemma 3.1 yields a rate of convergence of $p = 2$ for boxes Y for which f is differentiable for all $(x,r) \in Y$.

7.6 Some examples

In this section, we demonstrate the proposed circle detection model in three studies.

7.6.1 Detecting a single circle

In all our examples, the set $\mathscr{A}$ was calculated using Canny's edge detection algorithm as summarized in Section 7.3. Then, the corresponding circle detection problem with $\delta = 1$ and $C = 1$, see Equation (7.1), was solved using the geometric branch-and-bound algorithm.

The initial box X was defined as

$$X = [0, w] \times [0, h] \times [0, w+h],$$

where w is the width and h the height of the image. Note that if a lower bound $r_{\min}$ and an upper bound $r_{\max}$ for the circle to be detected are known, the initial box should be modified to

$$X = [0, w] \times [0, h] \times [r_{\min}, r_{\max}].$$

Moreover, for all subboxes $Y \subset X$ we applied the natural interval bounding operation as well as the bounding operation as given in Theorem 7.1 and we used an accuracy of $\varepsilon = 10^{-8}$.

In our examples shown in Figure 7.2, the circles were detected in a few seconds of computer time. Note that using only the natural interval bounding operation no problem was solved even for an accuracy of $\varepsilon = 10^{-4}$ in a time limit of one day. However, for some larger boxes $Y \subset X$ the natural interval bounding operation yields sharper bounds compared to the bounding operation as given in Theorem 7.1.

7.6.2 Detecting several circles

The presented method is also suitable for detecting several circles iteratively as follows. Once a circle is found, all demand points $a \in \mathscr{A}$ close to the circumference of that circle are deleted from $\mathscr{A}$. Then the algorithm starts again with the reduced set $\mathscr{A}$ and so on.

To be more precise, consider a distance threshold $d_{\min}$ and a termination value of $S_{\min}$. Then the following algorithm finds several circles where $\mathscr{A}$ is the input set found by Canny's edge detection method as before.

1. Set $k = 1$ and $\mathscr{A}_k = \mathscr{A}$.
2. Solve the circle detection problem for $\mathscr{A}_k$, let (x_k^*, r_k^*) be an optimal solution, and S_k^* be the optimal value of the objective function.
3. If $S_k^* > S_{\min}$ stop. Else set

$$\mathscr{A}_{k+1} = \mathscr{A}_k \setminus \left\{ a \in \mathscr{A}_k : \left| \|x_k^* - a\| - r_k^* \right| \leq d_{\min} \right\},$$

set $k = k+1$, and go to Step 2.

Two examples for the detection of several circles are given in Figure 7.3. For these examples we applied $S_{\min} = -20$ and $d_{\min} = 1$.

7.6.3 Impact of the penalty term

Finally, we discuss the impact of the penalty term introduced in Equation (7.1). Figure 7.4 shows an example where one circle should be found using the suggested algorithm. Here, the global optimization problem was solved twice. A first time with $C = 1$ and a second time with $C = 0$.

As can be seen in Figure 7.4, we do not obtain a suitable result using $C = 0$; that is the penalty term C should also be regularized for successful applications of the circle detection problem.

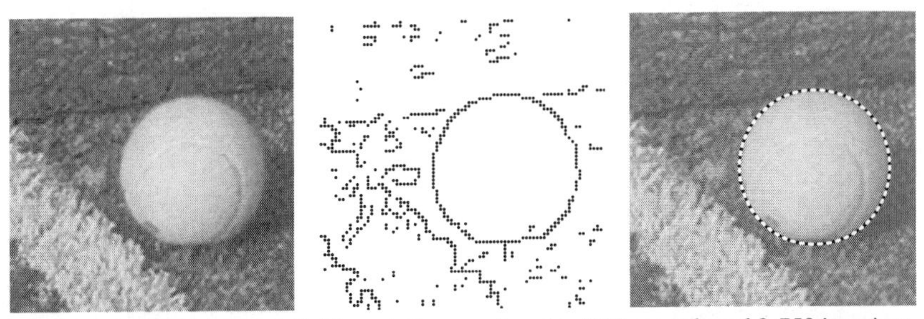

(a) Tennis ball with $|\mathscr{A}| = 640$. The solution was found in 7.35 seconds and 2,750 iterations.

(b) Heat and power station with $|\mathscr{A}| = 288$. The solution was found in 4.12 seconds and 2,727 iterations.

(c) Roundabout with $|\mathscr{A}| = 726$. The solution was found in 172.46 seconds and 31,210 iterations.

Fig. 7.2 Numerical results for the circle detection problem. In all examples, the original image, the set $\mathscr{A}$ found by Canny's edge detection algorithm, and the detected circle are shown.

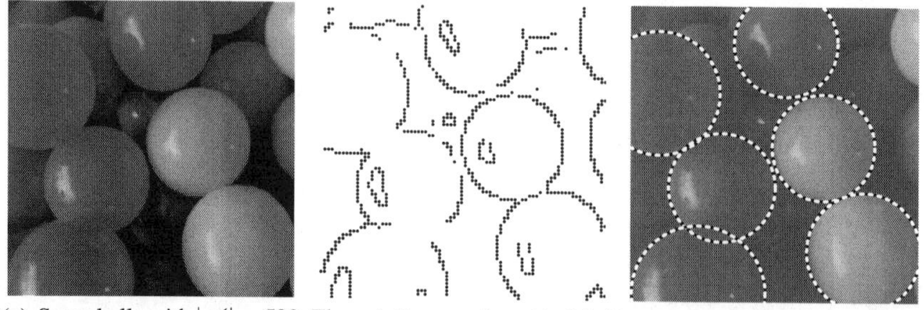

(a) Some balls with $|\mathscr{A}| = 599$. The solution was found in 88.41 seconds and 40,429 iterations.

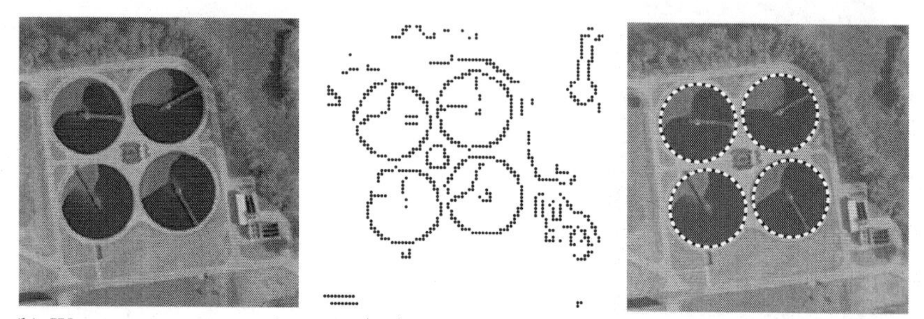

(b) Wastewater treatment plant with $|\mathscr{A}| = 682$. The solution was found in 203.4 seconds and 50,239 iterations.

Fig. 7.3 Numerical results for the circle detection problem finding several circles. In all examples, the original image, the original set $\mathscr{A}$ found by Canny's edge detection algorithm, and the detected circles are shown.

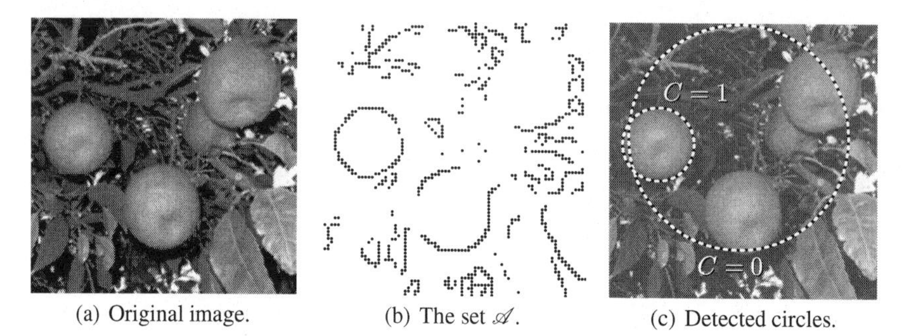

(a) Original image. (b) The set $\mathscr{A}$. (c) Detected circles.

Fig. 7.4 Numerical results to illustrate the impact of the penalty term for the circle detection problem.

Chapter 8
Integrated scheduling and location problems

Abstract As a second application of geometric branch-and-bound methods we present an integrated scheduling and location problem, namely the ScheLoc makespan problem. ScheLoc problems are location problems where we want to find an optimal location as well as an optimal schedule in an integrated model. Therefore, geometric branch-and-bound methods with mixed continuous and combinatorial variables are appropriate solution techniques for ScheLoc problems. In Section 8.1, we give a short introduction to ScheLoc problems before the ScheLoc makespan problem is presented explicitly in Section 8.2. In the following Sections 8.3 and 8.4, we show how to calculate lower bounds on the objective function as well as finite dominating sets for the combinatorial variables as required throughout the algorithm if we want to find an exact optimal solution. Some numerical experiences with comparisons to results reported in the literature are given in Section 8.5 before the chapter ends with a brief discussion in Section 8.6.

8.1 Introduction

For many applications in operations research it is obvious that dealing with problems in the usual sequential manner (taking the output of one problem as input of a next one) weakens the model and should be replaced by an integrated approach of solving the problems simultaneously. Integrated scheduling location problems, ***ScheLoc*** problems for short, are models for solving some scheduling and location problems simultaneously.

As an example, consider a container harbor where cargo has to be loaded on ships. In this application, one has to find positions for the ships on the berth (location) and to handle the corresponding loading process of the containers (scheduling). Traditionally, one first solves the location problem and then the scheduling problem. With the integrated approach we can find positions for the ships and define the loading process of the containers simultaneously.

Integrated scheduling location problems were first introduced by Hamacher and Hennes (2007) where machines could be located anywhere on a network. First results in the area of planar ScheLoc problems where machines can be located anywhere in a given planar region were established by Elvikis et al. (2009). Therein, some polynomial-time algorithms for the rectilinear norm and, more general, polyhedral gauges can be found. Using a geometric branch-and-bound algorithm, Kalsch and Drezner (2010) solved single-machine ScheLoc problems without pre-emption. Parts of the material presented here can also be found in Kalsch and Scholz (2010).

In the present chapter, our goal is to present an exact solution algorithm for the planar ScheLoc makespan problem as introduced in the following section. To this end, we apply the geometric branch-and-bound algorithm for mixed continuous and combinatorial optimization problems as introduced in Chaper 6.

8.2 The planar ScheLoc makespan problem

In the remainder of this chapter, we consider the planar ScheLoc makespan problem as studied in Elvikis et al. (2009) and Kalsch and Drezner (2010). For m given jobs consider a ***storage location*** $a_k \in \mathbb{R}^2$ and a nonnegative ***processing time*** p_k for $k = 1,\ldots,m$. All jobs must be scheduled on a single machine that can be placed anywhere on the plane $\mathbb{R}^2$. Hence, the general single-machine ScheLoc problem consists of choosing a machine location $x \in \mathbb{R}^2$ under the constraint that the set of jobs is completely processed and that all processing conditions are satisfied. Our goal is to optimize a scheduling objective function that depends not only on the sequence of jobs, but also on the choice of x.

To this end, all $k = 1,\ldots,m$ jobs are additionally characterized by the following parameters. The ***storage arrival time*** $\sigma_k \geq 0$ represents the time at which job k is available at its storage location a_k. After job k is available at its storage location a_k, we can start to move it from its storage to the machine. The ***travel speed*** $v_k > 0$ represents the rate of motion of job k, or equivalently the rate of change of position, expressed as distance per unit time. Hence, the time at which k can start its processing is dependent on its arrival time at the machine location x. Therefore,

$$r_k(x) \;:=\; \sigma_k + \tau_k \cdot \|x - a_k\|$$

is called the ***variable release date*** of job k, where $\|\cdot\|$ is a given norm and $\tau_k = 1/v_k$. Moreover, the sequence in which the jobs are to be processed on the machine is defined by a permutation π of $\{1,\ldots,m\}$. The set of all permutations of $\{1,\ldots,m\}$ is denoted by Π_m. Then for each sequence $\pi = (\pi_1,\ldots,\pi_m) \in \Pi_m$ and each machine location $x \in \mathbb{R}^2$, we can calculate the ***completion times*** for all jobs using the following recursion.

$$\begin{aligned} C_{\pi_1}(x) &= r_{\pi_1}(x) + p_{\pi_1}, \\ C_{\pi_k}(x) &= \max\left\{r_{\pi_k}(x), C_{\pi_{k-1}}(x)\right\} + p_{\pi_k} \end{aligned}$$

for all $k = 2,\ldots,m$, where p_{π_k} defines the processing time of job π_k. These completion times can be explicitly written by

$$C_{\pi_k}(x) = \max\left\{ r_{\pi_k}(x) + p_{\pi_k},\ r_{\pi_{k-1}}(x) + \sum_{j=k-1}^{k} p_{\pi_j},\ \ldots,\ r_{\pi_1}(x) + \sum_{j=1}^{k} p_{\pi_j} \right\}.$$

Thus, the maximum completion time or ***makespan*** in $x \in \mathbb{R}^2$ and for $\pi \in \Pi_m$ is given by

$$C_{\max}(x) = \max\{C_1(x),\ldots,C_m(x)\} = C_{\pi_m}(x). \tag{8.1}$$

Because the makespan should be minimized, we finally end up with the ***ScheLoc makespan problem***

$$\min_{\substack{x\in\mathbb{R}^2\\ \pi\in\Pi_m}} f(x,\pi),$$

where the objective function is given by

$$\begin{aligned} f(x,\pi) = f(x,\pi_1,\ldots,\pi_m) &= C_{\pi_m}(x) \\ &= \max\left\{ r_{\pi_m}(x) + p_{\pi_m},\ r_{\pi_{m-1}}(x) + \sum_{j=m-1}^{m} p_{\pi_j},\ \ldots,\ r_{\pi_1}(x) + \sum_{j=1}^{m} p_{\pi_j} \right\}. \end{aligned}$$

8.2.1 Fixed location

For any fixed machine location $x \in \mathbb{R}^2$, the ScheLoc makespan problem reduces to a classical makespan problem with fixed release dates $r_k = r_k(x)$ for all $k = 1,\ldots,m$ derived from the fixed machine location. In this case, we can use the well-known ***earliest release date*** rule, ERD rule for short; that is sequence the jobs in order of nondecreasing release dates to obtain an optimal job sequence; see Blazewicz et al. (2007), Brucker (2007), or Lawler (1973). Thus, for any machine location $x \in \mathbb{R}^2$ an optimal job sequence is given by a $\pi \in \Pi_m$ such that

$$r_{\pi_1}(x) \leq \cdots \leq r_{\pi_m}(x).$$

Such a π can easily be found by sorting $r_1(x)$ to $r_m(x)$.

8.2.2 Fixed permutation

For any fixed schedule $\pi \in \Pi_m$, the problem reduces to the weighted center location problem with addends. In particular, we have to minimize $g : \mathbb{R}^2 \to \mathbb{R}$ with

$$g(x) = \max\{\tau_1 \cdot \|x - a_1\| + c_1,\ \ldots,\ \tau_m \cdot \|x - a_m\| + c_m\} \tag{8.2}$$

and for $k = 1,\dots,m$ we obtain the addends

$$c_{\pi_k} = \sigma_{\pi_k} + \sum_{j=k}^{m} p_{\pi_j}.$$

If we consider, for instance, the rectilinear norm, this problem can be solved by the algorithm presented in Francis et al. (1992) in $\mathcal{O}(m^2)$ time. However, note that the problem can be transformed into a linear program with a fixed number of variables because the rectilinear norm can be represented by the maximum of four affine linear functions. Thus, we can solve the problem in linear time $\mathcal{O}(m)$; see Megiddo and Supowit (1984).

8.3 Mixed bounding operation

Before the calculation of some lower bounds is presented, the following result yields a dominating location set: we obtain a compact box X that contains an optimal solution for the continuous variables.

Lemma 8.1. *Consider the convex hull* $\text{conv}(a_1,\dots,a_m)$ *of the storage locations* $a_1,\dots,a_m$ *and let* $X \subset \mathbb{R}^2$ *be a box with*

$$\text{conv}(a_1,\dots,a_m) \subset X.$$

Then there exists an optimal solution (x^*,π^*) *for the ScheLoc makespan problem using the distance measure* $\|\cdot\| = \ell_p$ *for* $1 \le p \le \infty$ *such that* $x^* \in X$.

Proof. See Kalsch and Scholz (2010). Therein, a discussion about some more general distance measures can also be found. □

To obtain a lower bound for the ScheLoc makespan problem we can use the mixed location bounding operation as discussed in Section 6.3. To be more precise, we assume that

$$d_k^{\min}(Y) = \min\{\|x - a_k\| : x \in Y\}$$

can easily be calculated for $k = 1,\dots,m$. Because $\sigma_k \ge 0$ and $\tau_k > 0$, we can calculate lower bounds on the variable release dates in Y as follows:

$$r_k^{\min}(Y) = \min\{r_k(x) : x \in Y\} = \sigma_k + \tau_k \cdot d_k^{\min}(Y).$$

Now, determine a sequence $\mu \in \Pi_m$ using the ERD rule for $r_1^{\min}(Y),\dots,r_m^{\min}(Y)$; that is select $\mu \in \Pi_m$ such that

$$r_{\mu_1}^{\min}(Y) \le \cdots \le r_{\mu_m}^{\min}(Y).$$

Then, we determine the lower bound on the minimum makespan in Y as follows:

$$LB(Y) = \max\left\{ r^{\min}_{\mu_m}(Y) + p_{\mu_m}, r^{\min}_{\mu_{m-1}}(Y) + \sum_{j=m-1}^{m} p_{\mu_j}, \dots, r^{\min}_{\mu_1}(Y) + \sum_{j=1}^{m} p_{\mu_j} \right\}.$$

Theorem 8.1. *Let $Y \subset X$ be an arbitrary box. Then we have*

$$LB(Y) \leq \max\left\{ r^{\min}_{\pi_m}(Y) + p_{\pi_m}, \dots, r^{\min}_{\pi_1}(Y) + \sum_{j=1}^{m} p_{\pi_j} \right\} \leq C_{\pi_m}(x)$$

for all $x \in Y$ and all $\pi \in \Pi_m$.

Proof. See Kalsch and Drezner (2010). □

To sum up, we found the following mixed bounding operation for the ScheLoc makespan problem.

$$\kappa(Y) = (\kappa_1(Y), \dots, \kappa_m(Y)) \in \left\{ \mu \in \Pi_m : r^{\min}_{\mu_1}(Y) \leq \dots \leq r^{\min}_{\mu_m}(Y) \right\}, \quad (8.3)$$

$$LB(Y) := \max\left\{ r^{\min}_{\kappa_m(Y)}(Y) + p_{\kappa_m(Y)}, \dots, r^{\min}_{\kappa_1(Y)}(Y) + \sum_{j=1}^{m} p_{\kappa_j(Y)} \right\}, \quad (8.4)$$

$$r(Y) := c(Y). \quad (8.5)$$

Theorem 8.2. *The mixed bounding operation for the planar ScheLoc makespan problem as defined in* (8.3) *to* (8.5) *has a rate of convergence of $p = 1$.*

Proof. The presented bounding operation is an application of the general mixed location bounding operation as introduced in Subsection 6.3.3. Hence, the rate of convergence of $p = 1$ follows from Theorem 6.4. □

8.4 Dominating sets for combinatorial variables

For the ScheLoc makespan problem we find dominating sets for the combinatorial variables for all subboxes $Y \subset X$ such that we can enumerate all optimal solutions if $|\Omega(Y)|$ is small enough; see Chapter 6.

To this end, define again

$$d^{\min}_k(Y) = \min\{\|x - a_k\| : x \in Y\} \text{ and } d^{\max}_k(Y) = \max\{\|x - a_k\| : x \in Y\}$$

and furthermore

$$r^{\min}_k(Y) = \sigma_k + \tau_k \cdot d^{\min}_k(Y) \text{ and } r^{\max}_k(Y) = \sigma_k + \tau_k \cdot d^{\max}_k(Y).$$

Moreover, determine a sequence μ using the ScheLoc ERD rule for

$$r^{\min}_1(Y), \dots, r^{\min}_m(Y)$$

as before: select $\mu \in \Pi_m$ such that

$$r^{\min}_{\mu_1}(Y) \leq \cdots \leq r^{\min}_{\mu_m}(Y).$$

Then we define

$$\begin{aligned}\Omega_k(Y) = \{\mu_k\} \cup \{\mu_s : r^{\max}_{\mu_s} \geq r^{\min}_{\mu_k}, \ s = 1, \ldots, k-1\} \\ \cup \{\mu_s : r^{\min}_{\mu_s} \leq r^{\max}_{\mu_k}, \ s = k+1, \ldots, m\}\end{aligned}$$

for $k = 1, \ldots, m$ and set

$$\Omega(Y) = \Pi_m \cap (\Omega_1(Y) \times \cdots \times \Omega_m(Y)).$$

Example 8.1. As an example, consider the values of $r^{\min}_k(Y)$ and $r^{\max}_k(Y)$ for $k = 1, \ldots, 7$ as depicted in Figure 8.1.

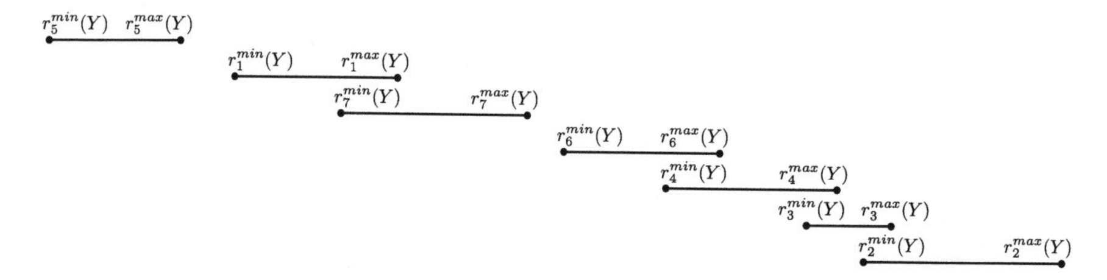

Fig. 8.1 Example for the calculation of the dominating sets for the combinatorial variables.

In this case, we have $\mu = (5, 1, 7, 6, 4, 3, 2)$ and we obtain the following sets.

$$\begin{array}{llll}\Omega_1 = \{5\}, & \Omega_2 = \{1,7\}, & \Omega_3 = \{1,7\}, & \Omega_4 = \{4,6\}, \\ \Omega_5 = \{3,4,6\}, & \Omega_6 = \{2,3,4\}, & \Omega_7 = \{2,3\}. & \end{array}$$

Summarizing, the ERD rule yields the following result.

Lemma 8.2. *Consider the ScheLoc makespan problem with $f(x, \pi)$, let $Y \subset X$ be an arbitrary subbox, and assume that $(x^*, \pi^*) \in Y \times \Pi$ is an optimal solution for*

$$\min_{\substack{x \in Y \\ \pi \in \Pi_m}} f(x, \pi).$$

Then we have $\pi^* \in \Omega(Y)$.

Proof. By construction of $\Omega(Y)$, the proof is similar to the proofs of Lemmas 6.2 and 6.3 and is therefore omitted here. □

8.5 Numerical results

For some computational experience, we generated $10 \leq m \leq 1,000$ storage locations with uniformly distributed parameters as shown in Table 8.1.

$a_k \in \{-500, -499, \ldots, 499, 500\}^2$
$\sigma_k \in \{0, 1, 2, \ldots, 25 \cdot m\}$
$v_k \in \{1, 2, \ldots, 10\}$
$p_k \in \{1, 2, \ldots, 50\}$

Table 8.1 Uniformly distributed parameters for $k = 1, \ldots, m$.

In our example instances, we used the rectilinear norm for all functions. Thus, as discussed in Section 8.2 the weighted center location problem (8.2) was solved by the algorithm presented in Francis et al. (1992).

Ten problems were solved for various values of m. As the initial box, we chose $X = [-500, 500]^2$, see Lemma 8.1, and for the ScheLoc ERD rule we implemented a Mergesort algorithm with a complexity of $\mathcal{O}(m \cdot \log m)$; see Knuth (1998).

To obtain an exact optimal solution, we applied a threshold for the cardinality of $|\Omega(Y)|$ of $M = 4$ for instances with $m < 1,000$ and $M = 8$ for instances with $m = 1,000$; see Chapter 6 and the numerical examples therein.

Our results are shown in Table 8.2. Therein, the runtimes, the number of iterations throughout the branch-and-bound algorithm, and the number of solved single facility location problems are reported.

	Runtime (Sec.)			Iterations			Location Problems		
m	Min	Max	Ave.	Min	Max	Ave.	Min	Max	Ave.
10	0.00	0.13	0.03	14	75	35.7	20	228	87.5
20	0.00	0.13	0.04	23	433	106.0	42	1,397	285.2
50	0.01	0.31	0.10	38	625	184.8	27	1,892	531.1
100	0.04	1.29	0.32	53	1,330	294.8	39	4,063	817.6
200	0.31	7.21	2.04	68	1,341	412.9	336	8,530	2,222.1
500	1.22	25.41	7.36	73	858	281.4	114	5,327	1,446.6
1,000	9.06	148.97	50.04	130	1,372	484.3	353	7,160	2,359.4

Table 8.2 Numerical results for the ScheLoc makespan problem.

All problem instances with up to $m = 200$ storage locations were solved in less than 10 seconds of computer time to an exact optimal solution. Even with $m = 1,000$, the average runtime was only 50 seconds. Note that using linear programming techniques Elvikis et al. (2009) could not solve instances of exactly the same problem with $m > 200$ to optimality within their limit of 24 hours of computer time on a machine comparable to the one we used. In contrast, our approach needed less than three minutes even for problem instances with $m = 1,000$ storage locations. A more detailed comparison of the runtimes using the benchmark instances introduced in Elvikis et al. (2009) is given in Table 8.3.

m	Geometric Branch-and-Bound (Sec.)	Elvikis et al. (2009) (Sec.)
50	0.21	983.24
100	0.37	5,219.70
150	0.44	31,504.00
200	4.77	85,032.00
250	2.46	not available
300	83.46	not available
350	29.96	not available
400	29.09	not available
450	7.85	not available
500	2.63	not available

Table 8.3 Comparison of the runtimes of our geometric branch-and-bound algorithm and the method in Elvikis et al. (2009) for ten given example instances with $50 \leq m \leq 500$ storage locations.

8.6 Discussion

In this chapter, we applied the geometric branch-and-bound algorithm for mixed continuous and combinatorial problems as a solution method for the ScheLoc makespan problem which provides an exact optimal solution. The presented numerical results show that the algorithm is fast even for large-scale instances with up to $m = 1,000$ storage locations. Furthermore, the branch-and-bound technique outperforms the linear programming approach as suggested in Elvikis et al. (2009).

Recall that the main profit of the geometric branch-and-bound algorithm is an exact solution. Therefore, further research ideas are exact solution algorithms for ScheLoc problems with pre-emtion as introduced in Kalsch and Scholz (2010).

However, it should be mentioned that the choice of the threshold M for the cardinality of $|\Omega(Y)|$ depends on the problem instance and the number of storage locations. Although we found exact optimal solutions in all of our problem instances, the algorithm might not terminate if M is too small; see Chapter 6. Therefore, we had to increase M in our numerical experiences for problem instances with $m = 1,000$ storage locations.

Chapter 9
The median line problem

Abstract As the last application considered in this text, we treat a facility location problem where we want to find an optimal location for a straight line. To be more precise, we want to locate a line in such a way that the sum of distances between that line and some given demand points is minimized. Although this problem is easy to solve in two dimensions, things become much more complicated in the three-dimensional case. Therefore, our aim is to apply the geometric branch-and-bound solution algorithm to the median line problem in the three-dimensional Euclidean space. To this end, after a short introduction and a literature review in Section 9.1, some theoretical results as well as a suitable problem formulation are given in Section 9.2. Furthermore, some bounding operations are suggested in Section 9.3 and numerical results can be found in Section 9.4.

9.1 Introduction

To present one more problem solved by the geometric branch-and-bound algorithm, we consider the ***median line problem*** in the three-dimensional Euclidean space: we seek a line that minimizes the sum of Euclidean distances to some demand points $\{a_1, \ldots, a_m\}$ in $\mathbb{R}^3$.

The median line problem in two dimensions with Euclidean distance and in the context of location theory was first analyzed by Wesolowsky (1975). Therein, it was shown that there exists an optimal line intersecting two data points which leads to a polynomial solution algorithm. Many generalizations such as general distance measures, line segments, and restrictions were studied in Morris and Verdini (1979), Morris and Norback (1980), Norback and Morris (1980), and Korneenko and Martini (1993) as well as in Schöbel (1999) and Díaz-Báñez et al. (2004) and references therein.

Although the Euclidean two-dimensional median line problem is well studied and exact polynomial-time algorithms are available, the three-dimensional problem becomes much harder and only a few references can be found in the literature. In

Brimberg et al. (2002), the authors discussed the problem of locating a vertical line as well as vertical line segments for any ℓ_p norm. It was shown that these problems can be essentially reduced to classical planar Weber problems. The work was extended in Brimberg et al. (2003). Therein, the three-dimensional median line problem was studied with some restrictions, for example, that all data points and/or the line to be located are contained in a given hyperplane. Furthermore, some heuristics for the general problem were presented, but without any numerical results.

Summarizing, the goal of the present chapter is to solve the general three-dimensional median line problem making use of the geometric branch-and-bound algorithm. To this end, we suggest a suitable problem formulation with only four variables and some bounding operations are given. Furthermore, it is shown how to find an initial box that contains at least one optimal solution. We remark that all results presented in this chapter can also be found in Blanquero et al. (2011).

9.2 Problem formulation

A line r in $\mathbb{R}^3$ has the form

$$r = r(x,d) = \{x+td \,:\, t \in \mathbb{R}\},$$

where $d \in \mathbb{R}^3 \setminus \{0\}$ is the direction of r and $x \in \mathbb{R}^3$. Moreover, we use the following notation.

Definition 9.1. For any $a \in \mathbb{R}^3$ and $x,d \in \mathbb{R}^3$ with $d \neq 0$ denote by

$$\delta_a(x,d) := \min_{t\in\mathbb{R}} \|x+td-a\|_2$$

the Euclidean distance from a to the line $r(x,d)$.

Note that we only consider the Euclidean distance in this chapter because the median line problem becomes much more complicated using other distance measures. For example, in general we cannot find a closed formula similar to the following one using the Euclidean norm.

Lemma 9.1. *Let $a \in \mathbb{R}^3$ and $x,d \in \mathbb{R}^3$ with $d \neq 0$. Then*

$$\delta_a(x,d) = \left\| x + \left(\frac{d^T(a-x)}{d^Td} \right) \cdot d - a \right\|_2 = \sqrt{\|x-a\|_2^2 - \frac{(d^T(a-x))^2}{d^Td}}. \quad (9.1)$$

Proof. Define the scalar function

$$g(t) := \|x+td-a\|_2^2.$$

Note that g is differentiable and strictly convex. Moreover, we have $g'(t^*) = 0$ for

$$t^* = \frac{d^T(a-x)}{d^T d}.$$

Hence, t^* minimizes g and we obtain $\delta_a(x,d) = \sqrt{g(t^*)}$. Furthermore, easy calculations lead to

$$((x-a)+t^*d)^T((x-a)+t^*d) = \|x-a\|_2^2 - \frac{(d^T(a-x))^2}{d^T d}$$

which proves the claim. □

In the remainder of this chapter our goal is to locate a line $r = r(x,d)$ in the three-dimensional Euclidean space that minimizes the sum of distances between r and a given set of demand points $a_1, \ldots, a_m \in \mathbb{R}^3$. Hence, the ***median line problem*** is given by

$$\min_{\substack{x,d\in\mathbb{R}^3\\ d\neq 0}} \sum_{k=1}^{m} \delta_{a_k}(x,d) = \min_{\substack{x,d\in\mathbb{R}^3\\ d\neq 0}} \sum_{k=1}^{m} \sqrt{\|x-a_k\|_2^2 - \frac{(d^T(a_k-x))^2}{d^T d}}. \tag{9.2}$$

9.2.1 Properties

Obviously, a line $r(x,d)$ is not uniquely defined by the pair (x,d). Indeed,

$$r(x,d) = r(x+\nu d, d)$$

for any $\nu \in \mathbb{R}$. Hence, we can assume without loss of generality that x is the intersection of r with the hyperplane

$$H_d = \{y : d^T y = 0\}. \tag{9.3}$$

Lemma 9.1 leads directly to the following corollary.

Corollary 9.1. *For any $a \in \mathbb{R}^3$ and $x,d \in \mathbb{R}^3$ with $d \neq 0$ and $d^T x = 0$ we have*

$$\delta_a(x,d) = \left\| x + \left(\frac{d^T a}{d^T d}\right)\cdot d - a \right\|_2 = \sqrt{\|x-a\|_2^2 - \frac{(d^T a)^2}{d^T d}}. \tag{9.4}$$

Moreover, we have $r(x,d) = r(x,\tau d)$ for any $\tau \in \mathbb{R}\setminus\{0\}$. Thus, we can also assume without loss of generality that $\|d\| = 1$. Hence, we can parameterize any line $r = r(x,d)$ by its associated pair (x,d) with $\|d\| = 1$ and $d^T x = 0$.

Corollary 9.2. *Let $a \in \mathbb{R}^3$ and $x,d \in \mathbb{R}^3$ with $\|d\| = 1$ and $d^T x = 0$. Then*

$$\delta_a(x,d) = \|x + d^T a \cdot d - a\|_2 = \sqrt{\|x-a\|_2^2 - (d^T a)^2}. \tag{9.5}$$

Next, let us consider the median line problem with fixed direction $d \in \mathbb{R}^3 \setminus \{0\}$ and the hyperplane H_d as defined in (9.3). We want to show that the median line problem with fixed d is equivalent to the planar Weber problem; see Subsection 1.3.2 or Drezner et al. (2001) for an overview. To this end, define the mapping

$$p_d : \mathbb{R}^3 \to H_d \text{ with } p_d(x) = x - \frac{d^T x}{d^T d} \cdot d$$

and note that $p_d(x)$ is the projection of x onto H_d.

Lemma 9.2. *Consider a fixed direction $d \in \mathbb{R}^3 \setminus \{0\}$. Then*

$$\delta_a(x,d) = \|p_d(x) - p_d(a)\|_2$$

for all $x, a \in \mathbb{R}^3$.

Proof. One has

$$\begin{aligned} \|p_d(x) - p_d(a)\|_2 &= \left\| \left(x - \frac{d^T x}{d^T d} \cdot d \right) - \left(a - \frac{d^T a}{d^T d} \cdot d \right) \right\|_2 \\ &= \left\| x + \left(\frac{d^T (a-x)}{d^T d} \right) \cdot d - a \right\|_2 = \delta_a(x,d) \end{aligned}$$

due to Lemma 9.1. □

We remark that the same result for the special case of a vertical line; that is for $d = (0,0,1)$, can also be found in Brimberg et al. (2002). Moreover, Lemma 9.2 leads directly to the following corollary which is a special case of the results in Martini (1994).

Corollary 9.3. *The (three-dimensional) median line problem with an arbitrary but fixed direction $d \in \mathbb{R}^3 \setminus \{0\}$ is equivalent to a (two-dimensional) Weber problem.*

To be more precise, for any $d \in \mathbb{R}^3 \setminus \{0\}$ one has

$$\min_{x \in \mathbb{R}^3} \sum_{k=1}^{m} \delta_{a_k}(x,d) = \min_{x \in \mathbb{R}^3} \sum_{k=1}^{m} \|p_d(x) - p_d(a_k)\|_2 = \min_{x \in H_d} \sum_{k=1}^{m} \|x - p_d(a_k)\|_2 . \tag{9.6}$$

The following basic property is important in order to restrict our search to a compact set.

Corollary 9.4. *An optimal solution $(x^*, d^*) \in \mathbb{R}^6$ exists for the median line problem such that the line $r = r(x^*, d^*)$ intersects the convex hull of A.*

Proof. Recall that for any fixed $d \in \mathbb{R}^3 \setminus \{0\}$ the median line problem is equivalent to a planar Weber problem; see Corollary 9.3.

Moreover, it is well known that there exists an optimal solution for the Weber problem which intersects the convex hull of the (projected) demand points

$$p_d(a_1), \ldots, p_d(a_m);$$

see, for instance, Drezner et al. (2001). Hence, there also exists an optimal line with fixed direction d which intersects the convex hull of A. This is true for any $d \in \mathbb{R}^3$ with $d \neq 0$, therefore the statement is shown. □

9.2.2 Problem parameterization

The six-dimensional problem, can be reduced to a four-dimensional problem in many ways. In the following we present the parameterization which turned out to be the most efficient one for the geometric branch-and-bound method.

Let $d = (d_1, d_2, d_3) \in \mathbb{R}^3$ and let us first assume that $d_3 = 1$ is fixed. We only need to consider $x = (x_1, x_2, x_3) \in \mathbb{R}^3$ such that $d^T x = 0$ as discussed at the beginning of this section. If we do so, we easily obtain

$$x_3 = -(x_1 d_1 + x_2 d_2).$$

With $a_k = (\alpha_k, \beta_k, \gamma_k)$ for $k = 1, \ldots, m$ and by making use of Corollary 9.1, we obtain in the case that $d_3 = 1$, the objective function,

$$f_3(x_1, x_2, d_1, d_2) := \sum_{k=1}^{m} \sqrt{g_3^k(x_1, x_2, d_1, d_2)},$$

where

$$\begin{aligned} g_3^k(x_1, x_2, d_1, d_2) := (x_1 - \alpha_k)^2 + (x_2 - \beta_k)^2 + (x_1 d_1 + x_2 d_2 + \gamma_k)^2 \\ - \frac{(d_1 \alpha_k + d_2 \beta_k + \gamma_k)^2}{d_1^2 + d_2^2 + 1}. \end{aligned}$$

In the same way we can also fix $d_1 = 1$ and $d_2 = 1$ which yield, renaming the four remaining variables always as x_1, x_2, d_1, and d_2,

$$f_1(x_1, x_2, d_1, d_2) := \sum_{k=1}^{m} \sqrt{g_1^k(x_1, x_2, d_1, d_2)},$$

$$f_2(x_1, x_2, d_1, d_2) := \sum_{k=1}^{m} \sqrt{g_2^k(x_1, x_2, d_1, d_2)},$$

where

$$g_1^k(x_1,x_2,d_1,d_2) := (x_1 d_1 + x_2 d_2 + \alpha_k)^2 + (x_1 - \beta_k)^2 + (x_2 - \gamma_k)^2 - \frac{(\alpha_k + d_1\beta_k + d_2\gamma_k)^2}{d_1^2 + d_2^2 + 1},$$

$$g_2^k(x_1,x_2,d_1,d_2) := (x_1 - \alpha_k)^2 + (x_1 d_1 + x_2 d_2 + \beta_k)^2 + (x_2 - \gamma_k)^2 - \frac{(d_1\alpha_k + \beta_k + d_2\gamma_k)^2}{d_1^2 + d_2^2 + 1}.$$

To sum up, the six-dimensional problem (9.2) is equivalent to the four-dimension problem

$$\min_{x_1,x_2,d_1,d_2 \in \mathbb{R}} f(x_1,x_2,d_1,d_2) \tag{9.7}$$

with

$$f(x_1,x_2,d_1,d_2) := \min\{f_1(x_1,x_2,d_1,d_2),\ f_2(x_1,x_2,d_1,d_2),\ f_3(x_1,x_2,d_1,d_2)\}.$$

9.3 Bounding operation and initial box

Recall that for any subbox

$$Y = X_1 \times X_2 \times D_1 \times D_2 \subset \mathbb{R}^4$$

we want to find a lower bound on the median line objective function

$$f(x_1,x_2,d_1,d_2) = \min\{f_1(x_1,x_2,d_1,d_2),\ f_2(x_1,x_2,d_1,d_2),\ f_3(x_1,x_2,d_1,d_2)\},$$

where

$$f_i(x_1,x_2,d_1,d_2) = \sum_{k=1}^{m} \sqrt{g_i^k(x_1,x_2,d_1,d_2)} \text{ for } i = 1,2,3$$

as defined before.

One obtains a first lower bound for this problem using the natural interval extension:

$$LB_1(Y) := F(Y)^L, \tag{9.8}$$

where $F(Y) = F(X_1,X_2,D_1,D_2)$ is the natural interval extension of $f(x_1,x_2,d_1,d_2)$; see Section 3.6.

For a second lower bound, we make use of the general bounding operation of order two as follows; see Section 3.5. Note that for $i = 1,2,3$ and $k = 1,\ldots,m$ the functions g_i^k are differentiable and define the linear functions

$$z_i^k(x_1,x_2,d_1,d_2) := g_i^k(\ell) + L_i^k(Y)^T \cdot \Big((x_1,x_2,d_1,d_2) - \ell\Big) \le g_i^k(x_1,x_2,d_1,d_2),$$

where $\ell = \ell(Y) = (X_1^L, X_2^L, D_1^L, D_2^L)$ is the left endpoint of Y and $L_i^k(Y)$ is a lower bound on the gradient $\nabla g_i^k(x_1,x_2,d_1,d_2)$ found by the natural interval bounding operation; see Section 3.6. Denoting by $V(Y)$ the 16 vertices of Y, these definitions yield

$$M_i^k(Y) := \min_{v \in V(Y)} z_i^k(v) = \min_{x \in Y} z_i^k(x).$$

Hence, we obtain the following result.

Lemma 9.3. *For $i = 1,2,3$ and $k = 1,\ldots,m$, the functions*

$$h_i^k(x_1,x_2,d_1,d_2) := \begin{cases} \sqrt{z_i^k(x_1,x_2,d_1,d_2)} & \text{if } M_i^k(Y) \ge 0 \\ 0 & \text{if } M_i^k(Y) < 0 \end{cases}$$

are concave on Y and satisfy

$$h_i^k(x_1,x_2,d_1,d_2) \le \sqrt{g_i^k(x_1,x_2,d_1,d_2)}$$

for all $(x_1,x_2,d_1,d_2) \in Y$.

Proof. Obviously, 0 is a concave function. Furthermore, if $M_i^k(Y) \ge 0$ then

$$z_i^k(x_1,x_2,d_1,d_2) \ge 0 \text{ for all } (x_1,x_2,d_1,d_2) \in Y$$

because z_i^k is linear. Moreover, the scalar function $u(t) = \sqrt{t}$ is concave and monotone increasing for $t \ge 0$, therefore we also know that

$$u(z_i^k(x_1,x_2,d_1,d_2))$$

is concave. Finally,

$$z_i^k(x_1,x_2,d_1,d_2) \le g_i^k(x_1,x_2,d_1,d_2) \text{ for all } (x_1,x_2,d_1,d_2) \in Y$$

and u is monotone increasing, therefore we know that

$$0 \le h_i^k(x_1,x_2,d_1,d_2) \le \sqrt{g_i^k(x_1,x_2,d_1,d_2)}$$

which proves the claim. □

With the help of Lemma 9.3 we obtain the following lower bound for the median line problem.

Theorem 9.1. *Define the functions*

$$h_i(x_1,x_2,d_1,d_2) := \sum_{k=1}^{m} h_i^k(x_1,x_2,d_1,d_2)$$

for $i = 1,2,3$ *and let*

$$h(x_1,x_2,d_1,d_2) := \min\{h_1(x_1,x_2,d_1,d_2),\ h_2(x_1,x_2,d_1,d_2),\ h_3(x_1,x_2,d_1,d_2)\}.$$

Then

$$LB_2(Y) := \min_{v\in V(Y)} h(v) \tag{9.9}$$

is a lower bound for the median line problem.

Proof. By Lemma 9.3, the functions h_i are concave for $i = 1,2,3$ and because the minimum of concave functions is concave again, h is concave on Y and we therefore obtain

$$\min_{x\in Y} h(x) = \min_{v\in V(Y)} h(v).$$

Lemma 9.3 furthermore states that

$$h(x_1,x_2,d_1,d_2) \le f(x_1,x_2,d_1,d_2) \text{ for all } (x_1,x_2,d_1,d_2) \in Y$$

and the theorem is shown. □

To sum up, we suggest using a combined bounding operation as follows.

Definition 9.2. A bounding operation for the median line location problem is given by

$$LB(Y) := \max\{LB_1(Y), LB_2(Y)\} \text{ and } r(Y) := c(Y).$$

We remark that other parameterizations of the median line problem are possible, for example spherical coordinates as suggested in Blanquero et al. (2009). We also implemented several other lower bounds using, for instance, techniques from d.c. programming or the centered interval bounding operation. However, all other parameterizations as well as all other lower bounds we tried were worse compared to the parameterization and the lower bounds presented here.

Finally, we need to find an initial box X that contains at least one optimal solution.

Theorem 9.2. *Without loss of generality assume that* $a_k \subset [-1,1]^3$ *for* $k = 1,\dots,m$. *Then the initial box*

$$X = [-\sqrt{3},\sqrt{3}] \times [-\sqrt{3},\sqrt{3}] \times [-1,1] \times [-1,1]$$

contains at least one optimal solution to the median line problem using the four-dimensional parameterization given in Equation (9.7).

Proof. Let $r(x,d)$ be an optimal solution to the median line problem with $x = (x_1,x_2,x_3)$ and $d = (d_1,d_2,d_3)$ such that $d^T x = 0$. According to Corollary 9.4 we can further assume that $r(x,d)$ intersects the convex hull of the demand points. Let us consider two cases.

1. Choose $s \in \{1,2,3\}$ such that $d_s = \max\{|d_1|,|d_2|,|d_3|\}$ and define

$$\tilde{d} = (\tilde{d}_1,\tilde{d}_2,\tilde{d}_3) = \frac{1}{d_s}\cdot(d_1,d_2,d_3).$$

   We obtain $\tilde{d}_s = 1$ and $|\tilde{d}_i| \le 1$ for $i = 1,2,3$. Because $r(x,d)$ and $r(x,\tilde{d})$ represent the same line, we have shown that there is an optimal solution (x_1,x_2,d_1,d_2) to the median line problem using the parameterization (9.7) such that $d_1,d_2 \in [-1,1]$.
2. Next, assume that $x_1 \notin [-\sqrt{3},\sqrt{3}]$ or $x_2 \notin [-\sqrt{3},\sqrt{3}]$. We know that $d^T x = 0$, hence the Euclidean distance from $0 \in \mathbb{R}^3$ to the line $r(x,d)$, see Corollary 9.1, is

$$\delta_0(x,d) = \|x\|_2 = \sqrt{x_1^2+x_2^2+(x_1d_1+x_2d_2)^2} > \sqrt{3}.$$

   However, $\|a\|_2 \le \sqrt{3}$ for all $a \in [-1,1]^3$, therefore the line $r(x,d)$ does not intersect the convex hull of the demand points, a contradiction.

To sum up, the initial box X contains at least one optimal solution to the median line problem. □

9.4 Numerical results

In this section we present some numerical experiences solving the median line problem. To this end, we employed the geometric branch-and-bound technique as well as the bounding operation presented in the previous section.

We randomly generated some demand points $a_k \in \{-1.0, -0.9, \ldots, 0.9, 1.0\}^3$ and all selected boxes were split into $s = 2$ congruent small subboxes; that is all selected boxes were bisected perpendicular to the direction of the maximum width component. Furthermore, as the initial box we chose

$$X = [-1.74,1.74] \times [-1.74,1.74] \times [-1,1] \times [-1,1];$$

see Theorem 9.2.

For various values of m, we solved 10 problem instances using $\varepsilon = 10^{-4}$. Our results are illustrated in Table 9.1. Therein, the minimum, maximum, and average runtimes as well as iterations throughout the branch-and-bound algorithm are reported. Moreover, Figure 9.1 shows the runtimes for all solved problem instances.

It can be seen that our algorithm took roughly half an hour to solve problem instances with only $m = 40$ demand points. Furthermore, the standard deviation for all numbers of demand points is quite high. For example, although all instances with $m = 5$ demand points could be solved in a few seconds of computer time, there was one instance that was solved in more than eight minutes. Similar observations can be done for other values of m. However, as shown in Blanquero et al. (2011),

	Runtime (Sec.)			Iterations		
m	Min	Max	Ave.	Min	Max	Ave.
5	4.92	493.23	71.21	68,494	1,623,368	289,066.2
10	30.57	2,896.46	377.65	98,107	1,745,858	333,412.3
15	65.17	2,354.88	717.47	148,055	1,375,130	492,475.6
20	297.37	1,275.09	665.25	218,439	716,939	438,584.1
25	125.80	1,304.55	468.25	116,882	1,007,635	387,641.2
30	92.58	1,026.36	495.95	105,472	936,559	425,608.1
35	123.61	1,977.49	603.99	107,864	1,113,556	438,750.9
40	893.49	3,864.53	1,767.02	427,091	2,079,367	900,580.4
45	217.49	4,664.80	1,635.44	166,098	1,902,454	816,639.9
50	397.64	6,235.79	2,089.75	189,778	1,935,559	784,289.4

Table 9.1 Numerical results for the median line problem with randomly generated input data and $\varepsilon = 10^{-4}$.

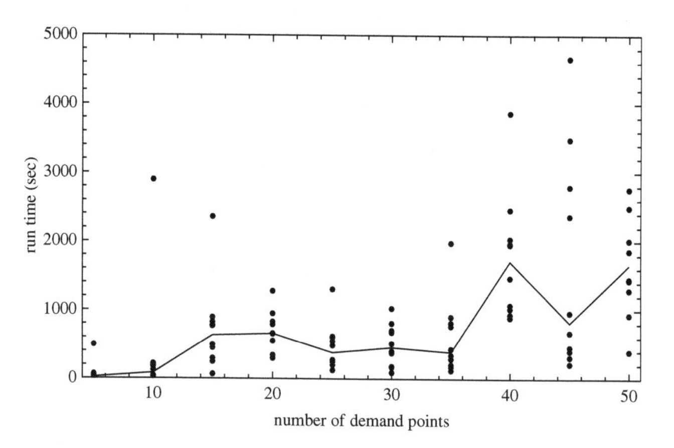

Fig. 9.1 Runtimes for all problem instances of the median line problem with randomly generated input data and $\varepsilon = 10^{-4}$. The polygonal chain represents the median of these values.

some more sophisticated data structures lead to a much faster implementation of the algorithms, but the number of iterations is of course still the same.

Furthermore, note that we only solved the unweighted median line problem in this chapter although the problem parameterization as well as the proposed lower bounds are still valid for weighted demand points with nonnegative weights and similar numerical results can be observed.

Example 9.1. Finally, we present a particular problem instance with $m = 50$ demand points as given in Table 9.2.

Using $\varepsilon = 10^{-6}$, we obtained after 976,861 iterations and a runtime of about 25 minutes the optimal line

(1.6, 0.2, 0.0)	(0.5, 0.4, 1.0)	(0.3, 1.8, 1.8)	(0.7, 1.4, 1.5)	(1.5, 1.8, 0.7)
(0.8, 2.0, 1.2)	(2.0, 1.8, 0.0)	(1.3, 0.6, 0.5)	(1.7, 0.1, 1.6)	(0.4, 1.4, 0.2)
(1.4, 1.2, 0.1)	(1.7, 0.3, 1.2)	(0.7, 2.0, 1.1)	(0.8, 1.2, 0.8)	(1.6, 1.7, 0.8)
(0.1, 1.5, 0.2)	(1.9, 0.6, 1.6)	(1.9, 0.9, 1.0)	(2.0, 0.2, 0.1)	(2.0, 0.6, 1.2)
(0.0, 0.4, 0.8)	(1.6, 1.0, 0.8)	(0.7, 1.0, 2.0)	(1.7, 0.1, 1.9)	(0.3, 1.5, 1.1)
(1.0, 1.9, 1.4)	(0.5, 1.5, 0.9)	(0.4, 0.7, 1.1)	(0.8, 0.9, 2.0)	(1.9, 0.2, 1.6)
(0.8, 1.3, 1.4)	(1.8, 1.8, 0.6)	(1.5, 1.1, 1.6)	(0.3, 0.9, 2.0)	(0.8, 0.1, 2.0)
(0.8, 1.1, 0.3)	(2.0, 1.8, 1.6)	(1.6, 1.5, 0.8)	(0.2, 2.0, 1.2)	(1.2, 1.6, 0.7)
(1.8, 1.4, 1.8)	(0.1, 1.2, 1.1)	(1.1, 0.3, 0.6)	(1.9, 1.4, 0.3)	(0.0, 0.9, 0.1)
(0.7, 1.5, 1.1)	(1.5, 1.2, 1.6)	(1.6, 0.0, 1.3)	(1.3, 1.7, 1.3)	(0.5, 0.0, 0.3)

Table 9.2 Demand points $a_1, \ldots, a_{50}$ for Example 9.1.

$$r = r(x^*, d^*) = \begin{pmatrix} 1.021705 \\ 1.173660 \\ 1.119308 \end{pmatrix} + t \cdot \begin{pmatrix} -0.980400 \\ 1.000000 \\ -0.153648 \end{pmatrix}$$

with an objective value of 36.893231.

Chapter 10
Summary and discussion

Abstract We conclude this book with a summary of the previous chapters, a discussion of the presented algorithms and techniques, and an outlook on further research as well as possible extensions. We start with the summary in Section 10.1 before a discussion of the suggested methods in Section 10.2. Finally, the work ends with an overview of further research ideas in Section 10.3.

10.1 Summary

In this book, we analyzed one of the most important techniques in deterministic global optimization, namely geometric branch-and-bound methods. The prototype algorithm was presented in Chapter 2 and possible variations were discussed therein. However, throughout all these branch-and-bound approaches the main task is to calculate the required lower bounds. Therefore, we introduced the concept of the rate of convergence which leads to a general convergence theory; see again Chapter 2. The main contribution of the present work can be found in Chapter 3 where we calculated the rate of convergence for several bounding operations collected from the literature. Using these tools, we provided a general solution technique that can be applied to a wide range of global optimization problems.

We furthermore suggested several extensions of the geometric branch-and-bound prototype algorithm. In Chapter 4, we discussed an extension for multicriteria global optimization problems where again the bounding operations collected in Chapter 3 can be used. The idea of our approach was to obtain a set that consists of ε-Pareto optimal solutions and contains all Pareto optimal solutions. Moreover, if one only wants to find a sharp outer approximation of the set of all Pareto optimal solutions, we presented some further discarding tests based on necessary conditions for Pareto optimality in Chapter 5. Finally, an extension for mixed continuous and combinatorial optimization problems was suggested in Chapter 6.

In the third part of the present book, some applications of geometric branch-and-bound techniques were discussed. We started with a problem in computer science in Chapter 7, namely with the circle detection problem. We pointed out how to use deterministic global optimization to detect shapes such as lines, circles, and ellipses in images. In a number of examples it was shown that the method is very accurate. In a second application in Chapter 8 we applied the branch-and-bound method for mixed optimization problems to integrated scheduling and location problems. In some numerical examples, we succeeded in finding exact global optimal solutions in all problem instances. Finally, we solved the median line problem: we applied the geometric branch-and-bound method to find a line in the three-dimensional Euclidean space that minimizes the sum of distances between that line and some given demand points.

10.2 Discussion

As mentioned before, a good bounding operation for a given global optimization problem is the most important choice for geometric branch-and-bound methods. Although several general bounding operations can be found in Chapter 3, it of course depends on the given problem which bounding operation should be preferred. To give an example, for the ScheLoc makespan problem even the location bounding operation with a rate of convergence of $p = 1$ yields sharp bounds not only for the rectilinear norm but also for the Euclidean norm as shown in Kalsch and Drezner (2010) and Kalsch and Scholz (2010). However, in several other location problems bounding operations with a rate of convergence of $p = 1$ are not suitable because in particular the bounds for smaller boxes are not sharp enough; see, for instance, Drezner and Suzuki (2004).

Furthermore, as mentioned in Section 1.4, a d.c. decomposition is never unique. Hence, the d.c. bounding operation of course depends on the chosen decomposition. For example, in the numerical results in Chapter 3 we used a d.c. decomposition that did not yield very good results although we have a rate of convergence of $p = 2$. With the help of Lemma 1.11, it might have also been possible to derive a particular d.c. decomposition for every subbox Y such that in general much sharper bounds could be found.

Moreover, in Section 3.5 we derived a general bounding operation with an arbitrary rate of convergence of p. However, to obtain a lower bound for a box Y we have to minimize a polynomial of degree $p-1$ over Y. Therefore, the general bounding operation becomes very intensive in runtimes even for $p = 3$ and $n > 2$. Thus, from the practical point of view the general bounding operation might not be suitable for $p > 2$.

To conclude the discussion about bounding operations, finally we remark that the number of iterations throughout the branch-and-bound algorithm does not only depend on the rate of convergence but also on the constant C and the optimization problem; see Example 10.1. Hence, for example, a bounding operation with a rate of convergence of $p = 3$ can be worse compared to a bounding operation with a rate of convergence of $p = 2$.

Example 10.1. Consider the objective function $f : \mathbb{R}^n \to \mathbb{R}$ with $f(x) = 0$ for all $x \in \mathbb{R}^n$. We choose the d.c. bounding operation using the d.c. decomposition

$$f(x) = g(x) - h(x) = x^T x - x^T x.$$

Hence, with $c = c(Y)$ for any box $Y \subset \mathbb{R}^n$ we obtain

$$m(x) = c^T c + 2c^T (x - c) - x^T x = -(c - x)^T (c - x) = -\|c - x\|_2^2$$

which yields the lower bound

$$LB(Y) = \min_{x \in Y} m(x) = -\frac{1}{4} \cdot \delta(Y)^2.$$

Therefore, we find

$$f(r(Y)) - LB(Y) \leq \frac{1}{4} \cdot \delta(Y)^2$$

for all boxes $Y \subset \mathbb{R}^n$. Thus, we find a bounding operation with a rate of convergence of $p = 2$ and with the small constant $C = \frac{1}{4}$. However, the branch-and-bound algorithm using the d.c. bounding operation with the suggested d.c. decomposition needs the worst case number of iterations as calculated in Theorem 2.1 and Example 2.2.

To sum up, this example shows that even if we are making use of a bounding operation with a rate of convergence of $p = 2$ and with a small constant C, the algorithm might be very slow. Hence, it also depends mainly on the particular problem if the geometric branch-and-bound algorithm leads to an efficient solution method.

Next, note that we considered only unconstrained global optimization problems in this work. If some constraints are present, they can also be bounded using the bounding operations collected in Chapter 3 and similar branch-and-bound methods lead to global optimal solutions; see, for example, Tuy (1998) or Hansen (1992).

We should also remark that even with very good bounding operations the geometric branch-and-bound method is only suitable for problems in small dimensions n inasmuch as we have to split every box in 2^n subboxes if we want to halve the diameter of the boxes. Moreover, in most of the discussed bounding operations a function should be evaluated at the 2^n vertices of the box. To this end, the runtime of the branch-and-bound technique increases exponentially with the dimension n. We found out that several problems up to $n = 6$ can be solved in general in a reasonable amount of time; see, for instance, the multisource Weber problem with three new facilities in Chapter 6. However, even in smaller dimensions not every global optimization problem can be solved efficiently or the calculation of sharp lower bounds is a quite difficult task.

Finally, we want to discuss the choice of the point $r(Y)$ as required for bounding operations; see Definition 2.2. Although the choice of $r(Y)$ is quite important from the theoretical point of view as shown in Theorem 3.2 and Example 2.1, there is almost no influence on the runtimes of the algorithm. To be more precise, even with $r(Y) = c(Y)$ for all bounding operations we obtain almost the same numerical results: the choice of $r(Y)$ might not be important for the runtimes solving particular problem instances.

10.3 Further work

We conclude this book with an overview of possible further research topics. First of all note that several facility location problems under block norms or polyhedral gauges as distance measures are solved by finite dominating sets; that is a finite set containing at least one global optimal solution; see, for example, Carrizosa et al. (2000) or Schöbel (1999). However, as mentioned before, because the cardinality of these sets mostly increases very fast with the number of demand points m, the suggested mixed combinatorial branch-and-bound algorithm might be a useful approach inasmuch as we do not have to evaluate the objective function at all points of the finite dominating set.

Furthermore, in Chapter 7 we suggested an accurate technique for the circle detection problem in image processing. But compared to the Hough transform or Ransac method, the presented technique is not fast enough. Therefore, a combination of commonly used techniques in image processing with global optimization tools might be a fruitful further research idea.

As also mentioned in the previous section, it does not only depend on the rate of convergence if the geometric branch-and-bound method is a suitable solution technique, but also on the particular problem. Up to now it is not clear which properties the objective function should satisfy such that the algorithm is fast and exact. Some theoretical results concerning this topic are desirable and left for further work.

Moreover, the rate of convergence of $p = 2$ for the d.c. bounding operation does not depend on the employed d.c. decomposition $f = g - h$; see Theorem 3.4. On the other hand, the constant C of course depends on the chosen d.c. decomposition and the question about an optimal decomposition arises; see, for instance, Ferrer and Martínez-Legaz (2009) and references therein. From the theoretical point of view, Lemma 3.1 states that the absolute value of the eigenvalues of $D^2 g$ should be as small as possible. However, empirical studies should be done that might lead to an answer about the optimal d.c. decomposition.

References

M. Avriel, W.E. Diewert, S. Schaible, and I. Zang, 1987. *Generalized Concavity*. Springer, New York, 1st edition.

E. Baumann. 1988. Optimal centered forms. *BIT Numerical Mathematics*, **28**: 80–87.

L. Bello, R. Blanquero, and E. Carrizosa. 2010. On minimax-regret Huff location models. *Computers & Operations Research*, **38**: 90–97.

R. Blanquero and E. Carrizosa. 2002. A D.C. biobjective location model. *Journal of Global Optimization*, **23**: 139–154.

R. Blanquero and E. Carrizosa. 2009. Continuous location problems and big triangle small triangle: Constructing better bounds. *Journal of Global Optimization*, **45**: 389–402.

R. Blanquero, E. Carrizosa, and P. Hansen. 2009. Locating objects in the plane using global optimization techniques. *Mathematics of Operations Research*, **34**: 837–858.

R. Blanquero, E. Carrizosa, A. Schöbel, and D. Scholz. 2011. A global optimization procedure for the location of a median line in the three-dimensional space. *European Journal of Operational Research*, **215**: 14–20.

J. Blazewicz, K. Ecker, E. Pesch, G. Schmidt, and J. Weglarz, 2007. *Handbook on Scheduling: Models and Methods for Advanced Planning*. Springer, New York, 1st edition.

T.M. Breuel. 2003a. Implementation techniques for geometric branch-and-bound matching methods. *Computer Vision and Image Understanding*, **90**: 258–294.

T.M. Breuel. 2003b. On the use of interval arithmetic in geometric branch and bound algorithms. *Pattern Recognition Letters*, **24**: 1375–1384.

J. Brimberg, P. Hansen, and N. Mladenović. 2006. Decomposition strategies for large-scale continuous location-allocation problems. *IMA Journal of Management Mathematics*, **17**: 307–316.

J. Brimberg and H. Juel. 1998a. A bicriteria model for locating a semi-desirable facility in the plane. *European Journal of Operational Research*, **106**: 144–151.

J. Brimberg and H. Juel. 1998b. On locating a semi-desirable facility on the continuous plane. *International Transactions in Operational Research*, **5**: 59–66.

J. Brimberg, H. Juel, and A. Schöbel. 2002. Linear facility location in three dimensions - Models and solution methods. *Operations Research*, **50**: 1050–1057.

J. Brimberg, H. Juel, and A. Schöbel. 2003. Properties of three-dimensional median line location models. *Annals of Operations Research*, **122**: 71–85.

J. Brimberg, N. Mladenović, and S. Salhi. 2004. The multi-source Weber problem with constant opening cost. *Journal of the Operational Research Society*, **55**: 640–646.

P. Brucker, 2007. *Scheduling Algorithms*. Springer, Berlin, 5th edition.

J. Canny. 1986. A computational approach to edge detection. *IEEE Transactions on Pattern Analysis and Machine Intelligence*, **8**: 679–698.

E. Carrizosa, H.W. Hamacher, R. Klein, and S. Nickel. 2000. Solving nonconvex planar location problems by finite dominating sets. *Journal of Global Optimization*, **18**: 195–210.

E. Carrizosa and F. Plastria. 1999. Location of semi-obnoxious facilities. *Studies in Locational Analysis*, **12**: 1–27.

L.G. Casado, I. Garcíal, and T. Csendes. 2000. A new multisection technique in interval methods for global optimization. *Computing*, **65**: 263–269.

L.G. Chalmet, R.L. Francis, and A. Kolen. 1981. Finding efficient solutions for rectilinear distance location problems efficiently. *European Journal of Operational Research*, **6**: 117–124.

P.C. Chen, P. Hansen, B. Jaumard, and H. Tuy. 1998. Solution of the multisource Weber and conditional Weber problems by D.-C. programming. *Operations Research*, **46**: 548–562.

W. Chuba and W. Miller. 1972. Quadratic convergence in interval arithmetic. Part I. *BIT Numerical Mathematics*, **12**: 284–290.

J. Clarke, S. Carlsson, and A. Zisserman. 1996. Detecting and tracking linear features efficiently. *Proceedings of the British Machine Vision Conference*, **1**: 415–424.

A.E. Csallner and T. Csendes. 1996. The convergence speed of interval methods for global optimization. *Computers and Mathematics with Applications*, **31**: 173–178.

A.E. Csallner, T. Csendes, and M.C. Markót. 2000. Multisection in interval branch-and-bound methods for global optimization – I. Theoretical results. *Journal of Global Optimization*, **16**: 371–392.

N.O. Da Cunha and E. Polak. 1967. Constrained minimization under vector-valued criteria in finite dimensional spaces. *Journal of Mathematical Analysis and Applications*, **19**: 103–124.

J.M. Díaz-Báñez, J.A. Mesa, and A. Schöbel. 2004. Continuous location of dimensional structures. *European Journal of Operational Research*, **152**: 22–44.

T. Drezner and Z. Drezner. 2004. Finding the optimal solution to the Huff based competitive location model. *Computational Management Science*, **1**: 193–208.

T. Drezner and Z. Drezner. 2007. Equity models in planar location. *Computational Management Science*, **4**: 1–16.

Z. Drezner. 1984. The planar two-center and two-median problems. *Transportation Science*, **18**: 351–361.

Z. Drezner, 1995. *Facility Location: A Survey of Applications and Methods*. Springer, New York, 1st edition.

Z. Drezner. 2007. A general global optimization approach for solving location problems in the plane. *Journal of Global Optimization*, **37**: 305–319.

Z. Drezner and H.W. Hamacher, 2001. *Location Theory - Applications and Theory*. Springer, New York, 1st edition.

Z. Drezner, K. Klamroth, A. Schöbel, and G. Wesolowsky. 2001. The Weber problem. In Z. Drezner, H.W. Hamacher, editors, *Location Theory - Applications and Theory*, pages 1–36. Springer, New York.

Z. Drezner and S. Nickel. 2009a. Constructing a DC decomposition for ordered median problems. *Journal of Global Optimization*, **45**: 187–201.

Z. Drezner and S. Nickel. 2009b. Solving the ordered one-median problem in the plane. *European Journal of Operational Research*, **195**: 46–61.

Z. Drezner and A. Suzuki. 2004. The big triangle small triangle method for the solution of nonconvex facility location problems. *Operations Research*, **52**: 128–135.

R.O. Duda and P.E. Hart. 1972. Use of the Hough transformation to detect lines and curves in pictures. *Communications of the ACM*, **15**: 11–15.

M. Ehrgott, 2005. *Multicriteria Optimization*. Springer, Berlin, 2nd edition.

D. Elvikis, H.W. Hamacher, and M.T. Kalsch. 2009. Simultaneous scheduling and location (ScheLoc): The planar ScheLoc makespan problem. *Journal of Scheduling*, **12**: 361–374.

A. Engau and M.W. Wiecek. 2007a. Exact generation of epsilon-efficient solutions in multiple objective programming. *OR Spectrum*, **29**: 335–350.

A. Engau and M.W. Wiecek. 2007b. Generating epsilon-efficient solution in multiobjective programming. *European Journal of Operational Research*, **177**: 1566–1579.

J. Fernández, B. Pelegrín, F. Plastria, and B. Tóth. 2006. Reconciling franchisor and franchisee: A planar biobjective competitive location and design model. *Lecture Notes in Economics and Mathematical Systems*, **563**: 375–398.

J. Fernández, B. Pelegrín, F. Plastria, and B. Tóth. 2007a. Planar location and design of a new facility with inner and outer competition: An interval lexicographical-like solution procedure. *Networks and Spatial Economics*, **7**: 19–44.

J. Fernández, B. Pelegrín, F. Plastria, and B. Tóth. 2007b. Solving a Huff-like competitive location and design model for profit maximization in the plane. *European Journal of Operational Research*, **179**: 1274–1287.

J. Fernández and B. Tóth. 2007. Obtaining an outer approximation of the efficient set of nonlinear biobjective problems. *Journal of Global Optimization*, **38**: 315–331.

J. Fernández and B. Tóth. 2009. Obtaining the efficient set of nonlinear biobjective optimization problems via interval branch-and-bound methods. *Computational Optimization and Applications*, **42**: 393–419.

A. Ferrer. 2001. Representation of a polynomial function as a difference of convex polynomials, with an application. *Lecture Notes in Economics and Mathematical Systems*, **502**: 189–207.

A. Ferrer and J.E. Martínez-Legaz. 2009. Improving the efficiency of DC global optimization methods by improving the DC representation of the objective function. *Journal of Global Optimization*, **43**: 513–531.

M.A. Fischler and R.C. Bolles. 1981. Random sample consensus: a paradigm for model fitting with applications to image analysis and automated cartography. *Communications of the ACM*, **24**: 381–395.

R.L. Francis, L.F. McGinnis, and J.A. White, 1992. *Facility Layout and Location: An Analytical Approach*. Prentice Hall, Englewood Cliffs, 2nd edition.

N. Guil and E.L. Zapata. 1997. Lower order circle and ellipse Hough transform. *Pattern Recognition*, **30**: 1729–1744.

H.W. Hamacher, 1995. *Mathematische Verfahren der Planaren Standortplanung*. Vieweg Verlag, Braunschweig, 1st edition.

H.W. Hamacher and H. Hennes. 2007. Integrated scheduling and location models: Single machine makespan problems. *Studies in Locational Analysis*, **16**: 77–90.

H.W. Hamacher and S. Nickel. 1996. Multicriteria planar location problems. *European Journal of Operational Research*, **94**: 66–86.

E. Hansen, 1992. *Global Optimization Using Interval Analysis*. Marcel Dekker, New York, 1st edition.

P. Hansen and B. Jaumard. 1995. Lipschitz optimization. In R. Horst, P.M. Pardalos, editors, *Handbook of Global Optimization*, pages 407–493. Kluwer Academic, Dordrecht.

P. Hansen, D. Peeters, D. Richard, and J.F. Thisse. 1985. The minisum and minimax location problems revisited. *Operations Research*, **33**: 1251–1265.

P. Hansen and J.F. Thisse. 1981. The generalized Weber-Rawls problem. In J.P. Brans, editor, *Operations Research*, pages 487–495. North-Holland, Amsterdam.

J.B. Hiriart-Urruty and C. Lemaréchal, 2004. *Fundamentals of Convex Analysis*. Springer, Berlin, 1st edition.

R. Horst, P.M. Pardalos, and N.V. Thoai, 2000. *Introduction to Global Optimization*. Springer, Berlin, 2nd edition.

R. Horst and N.V. Thoai. 1999. DC programming: Overview. *Journal of Optimization Theory and Applications*, **103**: 1–43.

R. Horst and H. Tuy, 1996. *Global Optimization: Deterministic Approaches*. Springer, Berlin, 3rd edition.

K. Ichida and Y. Fujii. 1990. Multicriterion optimization using interval analysis. *Computing*, **44**: 47–57.

B. Jähne, 2002. *Digital Image Processing*. Springer, Berlin, 5th edition.

M.T. Kalsch and Z. Drezner. 2010. Solving scheduling and location problems in the plane simultaneously. *Computers & Operations Research*, **37**: 256–264.

M.T. Kalsch and D. Scholz. 2010. A global solution method for single machine scheduling-location models in the plane. Unpublished.

P. Kanniappan. 1983. Necessary conditions for optimality of nondifferentiable convex multiobjective programming. *Journal of Optimization Theory and Applications*, **40**: 167–174.

D.E. Knuth, 1998. *The Art of Computer Programming. Volume 3: Sorting and Searching*. Addison-Wesley Longman, Amsterdam, 2nd edition.

N.M. Korneenko and H. Martini. 1993. Hyperplane approximation and related topics. In J. Pach, editor, *New Trends in Discrete and Computational Geometry*, pages 135–162. Springer, New York.

R. Krawczyk and K. Nickel. 1982. Die zentrische Form in der Intervallarithmetik, ihre quadratische Konvergenz und ihre Inklusionsisotonie. *Computing*, **28**: 117–137.

E.L. Lawler. 1973. Optimal sequencing of a single machine subject to precedence constraints. *Management Science*, **19**: 544–546.

T. Lindeberg. 1998. Edge detection and ridge detection with automatic scale selection. *International Journal of Computer Vision*, **30**: 117–154.

P. Loridan. 1984. ε-solutions in vector minimization problems. *Journal of Optimization Theory and Applications*, **43**: 265–276.

R.F. Love, J.G. Morris, and G.O. Wesolowsky. *Facilities Location: Models and Methods*. North-Holland, New York.

A.A.K. Majumdar. 1997. Optimality conditions in differentiable multiobjective programming. *Journal of Optimization Theory and Applications*, **92**: 419–427.

M.C. Markót, T. Csendes, and A.E. Csallner. 1999. Multisection in interval branch-and-bound methods for global optimization – II. Numerical Tests. *Journal of Global Optimization*, **16**: 219–228.

H. Martini. 1994. Minsum k-flats of finite point sets in $\mathbb{R}^d$. *Studies in Locational Analysis*, **7**: 123–129.

L.F. McGinnis and J.A. White. 1978. A single facility rectilinear location problem with multiple criteria. *Transportation Science*, **12**: 217–231.

N. Megiddo and K.J. Supowit. 1984. On the complexity of some common geometric location problems. *SIAM Journal on Computing*, **13**: 182–196.

E. Melachrinoudis and Z. Xanthopulos. 2003. Semi-obnoxious single facility location in Euclidean space. *Computers & Operations Research*, **30**: 2191–2209.

K.M. Miettinen, 1999. *Nonlinear Multiobjective Optimization*. Kluwer Academic, Boston, 1st edition.

J.G. Morris and J.P. Norback. 1980. A simple approach to linear facility location. *Transportation Science*, **14**: 1–8.

J.G. Morris and W.A. Verdini. 1979. Minisum l_p distance location problems solved via a perturbed problem and Weiszfeld's algorithm. *Operations Research*, **27**: 1180–1188.

A. Neumaier, 1990. *Interval Methods for Systems of Equations*. Cambridge University Press, New York, 1st edition.

N. Neykov, P. Filzmoser, R. Dimova, and P. Neytchev. 2007. Robust fitting of mixtures using the trimmed likelihood estimator. *Computational Statistics & Data Analysis*, **52**: 299–308.

S. Nickel. 1997. Bicriteria and restricted 2-facility Weber problems. *Mathematical Methods of Operations Research*, **45**: 167–195.

S. Nickel and J. Puerto, 2005. *Location Theory. A Unified Approach*. Springer, Berlin, 1st edition.

S. Nickel, J. Puerto, A.M. Rodríguez-Chía, and A. Weißler. 1997. General continuous multicriteria location problems. Technical report, University of Kaiserslautern, Department of Mathematics.

J.P. Norback and J.G. Morris. 1980. Fitting hyperplanes by minimizing orthogonal deviations. *Mathematical Programming*, **19**: 102–105.

Y. Ohsawa. 2000. Bicriteria Euclidean location associated with maximin and minimax criteria. *Naval Research Logistics*, **47**: 581–592.

Y. Ohsawa, N. Ozaki, and F. Plastria. 2008. Equity-efficiency bicriteria location with squared Euclidean distances. *Operations Research*, **56**: 79–87.

Y. Ohsawa, F. Plastria, and K. Tamura. 2006. Euclidean push-pull partial covering problems. *Computers & Operations Research*, **33**: 3566–3582.

Y. Ohsawa and K. Tamura. 2003. Efficient location for a semi-obnoxious facility. *Annals of Operations Research*, **123**: 173–188.

F. Plastria. 1992. GBSSS: The generalized big square small square method for planar single-facility location. *European Journal of Operational Research*, **62**: 163–174.

F. Plastria and M. Elosmani. 2008. On the convergence of the Weiszfeld algorithm for continuous single facility location-allocation problems. *TOP*, **16**: 388–406.

H. Ratschek and J. Rokne, 1988. *New Computer Methods for Global Optimization*. Ellis Horwood, Chichester, England, 1st edition.

H. Ratschek and R.L. Voller. 1991. What can interval analysis do for global optimization? *Journal of Global Optimization*, **1**: 111–130.

R.T. Rockafellar, 1970. *Convex Analysis*. Princeton University Press, NJ, 1st edition.

D. Romero-Morales, E. Carrizosa, and E. Conde. 1997. Semi-obnoxious location models: A global optimization approach. *European Journal of Operational Research*, **102**: 295–301.

P.J. Rousseeuw and A.M. Leroy, 1987. *Robust Regression and Outlier Detection*. John Wiley & Sons, New York, 1st edition.

A. Schöbel, 1999. *Locating Lines and Hyperplanes. Theory and Algorithms*. Kluwer Academic, Dordrecht, 1st edition.

A. Schöbel and D. Scholz. 2010a. The big cube small cube solution method for multidimensional facility location problems. *Computers & Operations Research*, **37**: 115–122.

A. Schöbel and D. Scholz. 2010b. The theoretical and empirical rate of convergence for geometric branch-and-bound methods. *Journal of Global Optimization*, **48**: 473–495.

D. Scholz. 2010. The multicriteria big cube small cube method. *TOP*, **18**: 286–302.

D. Scholz. 2011a. General further discarding tests in geometric branch-and-bound methods for non-convex multicriteria optimization problems. *Journal of Multi-Criteria Decision Analysis*. DOI 10.1002/mcda.474.

D. Scholz. 2011b. Theoretical rate of convergence for interval inclusion functions. *Journal of Global Optimization*. DOI 10.1007/s10898-011-9735-9.

C. Singh. 1987. Optimality conditions in multiobjective differentiable programming. *Journal of Optimization Theory and Applications*, **53**: 115–123.

A.J.V. Skriver and K.A. Anderson. 2003. The bicriterion semi-obnoxious location (BSL) problem solved by an ε-approximation. *European Journal of Operational Research*, **146**: 517–528.

B. Tóth and T. Csendes. 2005. Empirical investigation of the convergence speed of inclusion functions in a global optimization context. *Reliable Computing*, **11**: 253–273.

B. Tóth, J. Fernández, B. Pelegrín, and F. Plastria. 2009. Sequential versus simultaneous approach in the location and design of two new facilities using planar Huff-like models. *Computers & Operations Research*, **36**: 1393–1405.

H. Tuy. 1996. A general D.C. approach to location problems. In C.A. Floudas, P.M. Pardalos, editors, *State of the Art in Gloabal Optimization: Computational Methods and Applications*, pages 413–432. Kluwer Academic, Dordrecht.

H. Tuy, 1998. *Convex Analysis and Global Optimization.* Kluwer Academic, Dordrecht, 1st edition.

H. Tuy, F. Al-Khayyal, and F. Zhou. 1995. A D.C. optimization method for single facility location problems. *Journal of Global Optimization*, **7**: 209–227.

H. Tuy and R. Horst. 1988. Convergence and restart in branch-and-bound algorithms for global optimization. Application to concave minimization and d.c. optimization problems. *Mathematical Programming*, **41**: 161–183.

E.V. Weiszfeld. 1937. Sur le point pour lequel la somme des distances de n points donné est minimum. *Tohoku Mathematical Journal*, **43**: 335–386.

G.O. Wesolowsky. 1975. Location of the median line for weighted points. *Environment and Planning A*, **7**: 163–170.

D.J. White. 1986. Epsilon efficiency. *Journal of Optimization Theory and Applications*, **49**: 319–337.

R.K.K. Yip, P.K.S. Tam, and D.N.K. Leung. 1992. Modification of hough transform for circles and ellipses detection using a 2-dimensional array. *Pattern Recognition*, **25**: 1007–1022.

M. Zaferanieh, H. Taghizadeh Kakhki, J. Brimberg, and G.O. Wesolowsky. 2008. A BSSS algorithm for the single facility location problem in two regions with different norm. *European Journal of Operational Research*, **190**: 79–89.

Index